SYSTEMIC ENERGY NETWORKS

SYSTEMIC ENERGY NETWORKS, Vol. 1.
The theory of Systemic Design applied to Energy sector.

by Silvia Barbero

foreword by Luigi Bistagnino

first published in 2012 by lulu.com

Copyright year: 2012
Copyright Note: Silvia Barbero. All rights reserved.
The information referred to above constitute this copyright notice: © 2012 Silvia Barbero. All rights reserved. Except for the quotation of short passages for the purposes of criticism and review, no part of this pubblication may be reproduced, stored in a retriva system, or trasmitted, in any form or any means, electronic, mechanical, photocopying, recording or otherwise, without the prior permission of the publisher.

ISBN 978-1-291-04436-2

General supervision: Silvia Barbero
Book and cover design: Amina Pereno
Graphics and editing: Amina Pereno

SILVIA BARBERO, PhD in Production Systems and Industrial Design at Politecnico di Torino, she is research fellow and lecturer about environment requirements of industrial products. She applies to research and didactics the individual involvement in ecodesign and environmental sustainability. Her speciality is the use of an integrated approach to coordinate new technologies, economic, social and environmental issues in the field of sustainable energy for distributed economies. In June 2010, she founded Plug Creativity, no-profit cultural association, that aims to connect ethics (social and environment commitment) with the graphics and creativity.

SYSTEMIC ENERGY NETWORKS

The theory of Systemic Design applied to Energy sector

foreword by Luigi Bistagnino

volume 1

Silvia Barbero

CONTENTS

7 Foreword by Luigi Bistagnino
9 Preface
11 Introduction
14 List of abbreviations

PART I RESEARCH FIELD
17 chapter 1: Energy sector
23 chapter 2: Overall status of the energy sector in EU

PART II RESEARCH METHODOLOGY
39 chapter 3: Systemic Design theory
63 chapter 4: Case Studies analysis: practice
83 chapter 5: Cross Analysis

PART III GOAL
101 chapter 6: Design and test framework to local economic development by Systemic Design
111 chapter 7: Conclusions

117 References

FOREWORD

In everyday life we do not mind the quality of energy we use and the connection with the whole system.

However, a new awareness, about **energy saving**, is consolidating and people become new behaviours concerning the proper use of both electricity and thermal power, also because of the increasing of electricity bill. The incandescent light bulbs have been replaced with luminescence ones and, in some cases, even with led ones; many household are no longer held on, and also television and hi-fi equipments are turned off, not in standby; during the winter, the temperature inside the buildings are no longer high, but always closer to the recommended 18 °C; and so on also for the other energy-consuming uses that were spread over. It is common to think that can be enough to improve good habits in the use. Sometimes, non-renewable resources are integrated with small domestic installations, as photovoltaic or solar thermal panels, but **we should think also about the whole system that we move with our choices**. In fact it is assumed to have the electricity or heat, but we are not totally aware about the consequences, if we use oil, probably our action is going to contribute a political system that we do not directly approve. The resource, we are using, comes from countries against which maybe we disagree deeply!

It's a simple example that calls us to be consistent in our actions.

Too often the action is not connected to the system that it really puts into action.

Fortunately, it is spreading greater awareness about issues that affect everyday life, people begin to peek out of the narrow confines of their personal life, perceiving that many others are thinking like us: **everyone realizes to be part of a big community**.

The relationships among individuals form groups that interact and diffuse their opinions; so it determines new approaches to actual problems. **A new consciousness re-connects everyday life with the territory in which people lives** and in which the quality would be restored. This awareness leads us to think about local environmental resources, nutrition and energy, realizing that these relationships give the quality and the main feature of the place where we live.

Then, why our actions should have a negative impact on the lives of others and their territory?

These reasons are the starting point of the work that is presented in this book in which the topic of energy is tackled starting from the analysis of local resources of the area in which people live. So we can discover new and correct ways of use, making decisions that take into account the local area, and we do not waste potentialities that belong to our community.

This close relationship creates a solid network of awareness that helps to naturally enhance the lives of individuals and the communities that they form together. We do not need bombastic technological innovations, **we simply have to change our point of view and everyday choices will get automatically changed**.

LUIGI BISTAGNINO

LUIGI BISTAGNINO, Architect and designer, lives and works in Torino, Italy.
He attends to eco-compatibility of industrial products and components. He designed objects actually in production and won national and international design prizes such as Il Compasso d'Oro, ADI.
Founder of the research group on Systemic Design, with the goal of developing products and processes in order to obtain zero emission.
Full Professor of Industrial Design, president of Industrial Design Courses at Politecnico di Torino, he is the author of essays and articles published on many important national and international reviews. His main publications, among the others, are: Systemic Design, Slow Food 2nd edition, Bra 2011; The outside shell seen from the inside, CEA, Milano 2008; Designpiemonte, Agit, Beinasco (Torino, Italy), 2007; Design with a future, Time&Mind, Torino, Italy, 2003; Ecodesign in the EU, The Kuopio Academy, Kuopio, Finland, 2000.

PREFACE

Local Economic Development has significance within the cultural context in which it operates and is currently becoming an increasingly important part in international cooperation. In the last decade, the possibility to enable scenarios of bottom-up economic and social development, led by local actors, is gaining ground in the last decade. In this context of steady change, the national and local governments, like businesses and other organizations must rethink development strategies in order to lead the change to widespread prosperity of the people and the ecosystem. To reach these goals, it needs to create **connections between local material, energetic and informational resources, generating multiple and complex solutions**. In particular, **energy** is a *common good*, considered as fundamental need and shared by the society, but also it is an *individual right* to increase our own capability. The energy affects the lives of individuals and their various combinations: the environmental impact of businesses is largely determined by external and internal exchanges of resources. In Systems Theory, the economic, social and environmental benefits are required through the efficiency of flows. The local energy guarantees autonomy and derives manly from renewable or perpetual sources.

The **theory of Systemic Design** and the **practice of Case studies** are tacked with the same keys: what, why, when, where and how. **The analysis of complex systems by scientific means** supports decision making, that approach turns environmental problems into business opportunities, so companies can transform unwanted waste into valuable products. The **cross analysis** of the case studies and of the theoretical/practical parts is crucial to find the main results for designing the framework. That analysis aims to turn the theoretical knowledge into pathways of change suitable to the needs and capacities specific to regions or localities, and to replicate this model of territorial energy nets. Rather than setting out with a proven "one-fits-all" solution in hand to catalyse the development of Systemic Design networks, this book primarily focuses on continuously fine-tuning and approach so as to make it best suited to dynamic region specific contexts.

In conclusion, the adoption of more **systemic production and person-centered approaches**, holds great potentiality to create **Local Economical Development**. The linkages between materials, energy, people and their knowledge are mapped out clearly, to design efficient path towards sustainable ways to use and re-use untapped resources. Using these territorial networks, we can collectively create energy systems that are beneficial for the people, the environment, and pleasurable for all. In the field of renewable energy the creation of **sustainable infrastructures and agile energy systems could help the development of territories**. Furthermore, the designed framework is tested in a micro-system in Italy to verify it with continuous feedbacks and to realize a concrete replicable pilot project. It is demonstrated that the green energy production in systemic nets of small and distributed plants helps the success and the sustainability of territories.

INTRODUCTION

The socio-economical growth is strictly connected with the access to a secure and affordable supply of energy, because with a permanent availability of energy in its various forms, especially electricity, people can learn, produce, share and increase their activities in wide meaning. But the emissions associated with energy generation and consumption is also central to number of key environmental issues. Therefore, the aim is to define sustainable models using a holistic theory that take into account not only the economical and environmental aspects, but also the social ones.

The research field is the energy sector, in particular the **production of green energy from renewable resources** (wood, biomass, waste), and not perpetual (sun, wind, water). This book will focus on the production even if it is stricktly linked to the consumption of it. The goal is to satisfy the actual need with a larger share of bioenergy. The economic evaluation of energy provides a **quantitative index** of its importance. Everyone knows that the price of oil influences global market, but only when we are faced with emergencies, such as the blackouts occurring throughout the world in the last years, we start to realize the **qualitative effects** of insufficient energy. Globalized society is touching the limit of its development and is becoming aware of its dependency on energy sources and its need to achieve sustainability for its consumption requirements in a complex scenario of social, economical, geopolitical influences.

The greatest energy demand is actually faced into two ways: by **increasing exploitation** of the planet's resources and by conducting a more intensive search for **energy efficiency**. These solutions, however, create rebound effects that **do not guarantee the long-term sustainable development** we are hoping for. The legitimate expectations of energy procurement are only valid in conditions of limit or sustainability, following the **ethical principles** of responsibility and precaution.

The European Union (EU)'s energy production has been declining in the last five years, where the largest share is the nuclear (30%), followed by solid fuels (22%), gas (20%), oil (14%) and renewable resources (14%), although the contribution of the latter is expected to increase significantly in the future. In any case, the current fuel mix varies widely in the EU Member States; to a certain degree it depends on the domestic resource/production pattern. The goals of **EU's New Energy Policy** are to reduce of 20% in Green-House Gas (GHG) emissions compared to 1990, along with the production of 20% share from renewables in the final energy demand by 2020, and to bring about a substantial improvement in energy efficiency[1]. **The focus of this research is in the second point: design a framework to encourage and represent the production of energy from renewable resources.**

---

[1] Climate Action – Energy for a Changing World, 23 January 2008.

The **research methodology** in this book is based on the **combination of theory and practice that resolves into a cross analysis**. The **theory of Systemic Design** and the **practice of case studies**, that are explained in details in another book [2], are tacked with the same key questions: *what, why, when, where* and *how*. Utilizing a **combination of desk and field research** lets on a more in-depth understanding of reality from different viewpoints, which is crucial when exploring topics or issues involving a range of actors, like the energy sector. This book is heavily focused on qualitative methods, including literature reviews, case studies, site visits, stakeholder interviews, industry interactions and international agencies connections. The foundation of this research is therefore the use of **diverse kinds of data sources** and a **mix of qualitative methods**. The *desk research* detects the information already written by others, in a important identification of the sources with their reliability. The *field research* directly observes the reality with empirical and personal experience. That means the *desk research* is less expensive and also less original; but it helps to define the headway processes of the research. The combination of *desk* and *field research* guarantees a reach and deep understanding of the facts to define an **original framework** (Celaschi & Deserti, 2007).

The **main result** is a **framework supporting the evolution of a new economy** with different sets of industrial relations, where long-term sustainability and success of interdependent activities are priorities over maximising economic growth or competitive advantage of individual entities. The projects based on renewable energy, teach how the creation of sustainable infrastructures and agile energy systems could develop a region. So the conclusion is that **green energy produced in small plants and distributed in the territory** favours the success and the sustainability. Reading in the right way the environment is possible to design the right technology that produce green energy and that is connected with other renewable resources. Such **agile system** can be a new paradigm for both energy efficiency and reliability for any region or country.

From the opening of the energy market, companies had new possibilities to design energy supply scheme and policy. Most companies have not developed yet technical and economical know-how to deal with distributed energy generation concept and are still concentrating mainly on their core-business and on short-coming economical performances. The **theoretical model was tested and improved with the practical experimentation in Agrindustria**, so the **designed framework is already verified and the pilot project is replicable**.

The development of **Distributed Energy Resources concept** is linked to the **territorial specific features**, both from natural resources and human activities point of view. This is currently not well established and, paradoxically, it can be noted that the liberalization of electricity market can favour the opposite situation of customers buying energy from a very distant and different territory, wasting the local potential of energy generation.

[2] Barbero, S. (2012) *Systemic Energy Networks. The practice of macro and micro case studies*. Raleigh, USA: Lulu Enterprises, Inc.

Local communities will benefit from implementation of the project directly through a rational use of energy resources and the enhancement of those type of renewable low environmental impact; indirectly because the implementation of that framework will vehicle a spreading of new culture and also the emergence of local initiatives relating to environmental issue. Finally, the **government** and the **municipalities** also will receive the results of the proposal because from the observation of socio-economic system generated from the implementation of this project, will allow local administrators to acquire the tools to make a set of guidelines for creating a plan for local sustainability in the medium term, and the creation of new normative instruments to respond to sustainability requirements. With the cooperation of **universities**, **companies** and **public bodies**, the exchange of proficiencies between expert local organizations and learning local organizations are facilitated. The goal is to institutionalize sustainable energy policies and implement sustainable energy action plans.

In conclusion, the adoption of more **systemic productions** and **person-centered approaches**, including more innovative capture and generation of energy holds great potential to create Local Economic Development. When the linkages between materials, energy, people and their knowledge are mapped out clearly, efficient pathways towards sustainable ways to use and reuse untapped resources will become apparent. Using these networks, energy systems can be collectively innovated, so they are beneficial and pleasurable for the people the environment. In the field of renewable energy the creation of sustainable infrastructures and agile energy systems could develop a region. **Green energy produced in small plants and distributed in territories helps the success and the sustainability**.

LIST OF ABBREVIATION

BAP	Biomass Action Plan
CAP	Common Agricultural Policy
EU	European Union
GHG	Green House Gases
LED	Local Economic Development
R&D	Research & Development
SD	Systemic Design
ZERI	Zero Emissions Research and Initiatives

PART I

RESEARCH FIELD

ENERGY SECTOR

Energy is one of the basic factors that determines the **competitiveness of a country's economy** and the **quality of life of its population**. Permanent availability of energy in its various forms, especially electricity, is required in today's globalised world.

The economic evaluation of energy provides a **quantitative** index of its importance. The price of oil influences the global market, but only when emergencies are faced, such as the blackouts which have occurred throughout the world in the past few years, do the **qualitative** effects of insufficient energy become evident. These crises are becoming stronger and more frequent due to the growing complexity of the economic system. The globalized society is aiming at the limits of its development and is becoming aware of its dependency on energy sources and on sustainability for its consumption requirements.

In this complex scenario three types of **actors** are involved and all of them have different relationships with energy (figure 1.0.I):

- **citizens**: energy as a commodity, is a consumer's right for self-improvement and should be safe for the environment and for people's health;

- **public administration**: energy as a public good, is a need (common resource) for society as a whole, and generates complex geopolitical strategies;

- **private companies**: energy as an asset, brings profit and contributes to economic growth.

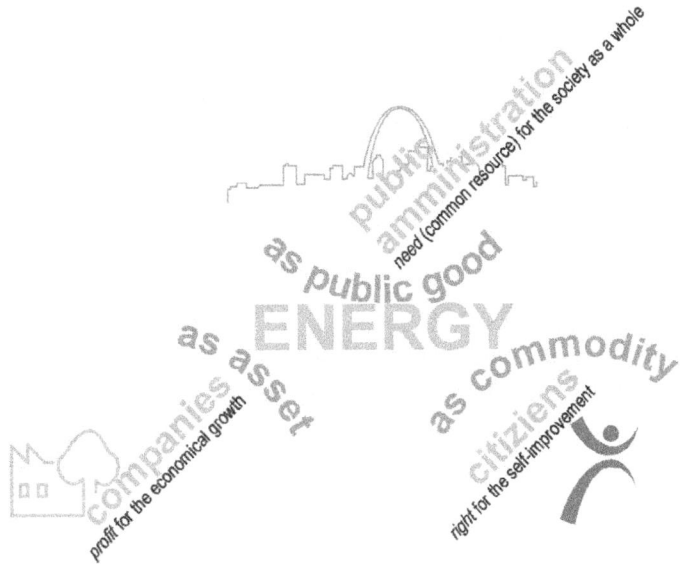

figure 1.0.I: energy as right/need/profit for the different actors

Of these three groups, the weakest are the citizens because they have only a few tools to influence the energy market, even if one of the biggest challenges for the future is the self- production of energy, as the economist Jeremy Rifkin outlines (Rifkin, 2002). The citizens consume energy, but they give it back in the form of social and collective intelligence. For that reason, the responsibility becomes a categorical imperative (Kant, 1785) concerning the ecosystem and its energy component.

The legitimate expectations of energy procurement are only valid in conditions of limit or sustainability, following the ethical principles of responsibility and precaution.

If we consider non-renewable energies (mainly oil and natural gas), the responsibility principle comes into play, by which the ethical decision follows infra-generational (towards present generations) and inter-generational (towards future generations) equity. Those resources are limited to natural deposits that do not follow geographical or political borders. So, nowadays, there are many problems related to the power that a few countries have over others just because they have the deposits. Furthermore, those deposits are not unlimited, so sooner or later[1] (it depends on the resources, on the technology used to extract them and on the previsions made by different scientists) they will run out. For sure, this is not fair for future generations that will need the same, or more, amount of energy without the current possibilities we have to use natural resources.

Energy with high environmental risk, like nuclear energy, brings the precaution principle, wich stipulates that given the reasonable doubt that an action or behaviour could cause seri- ous, and potentially fatal, damage to the community, even without evidence to substantiate this, proper measures to prevent it should be taken. That is the typical case of nuclear power stations and their safety not only for the environment, but for workers and the people that live in the surroundings.

A radically new approach to this issue is, therefore, necessary to lead the global civil society (Capra, 2002) to redesign the flow of resources and relationships within local context and generate multiple and complex solutions. In that sense renewable resources are one of the possible solutions that can be adopted, but to increase exponencially the benefit we can generate energy from waste, which in current society have continuous renewability. Never- theless, the principle of beneficiality states that actions should move towards benefit, and not only avoid harm (Sgreccia, 2007).

While looking at the ethical principles involved in the role of energy in current society, it can be said that democracy is the only model of government, which manages energy in a paradoxical way: it wants more and more energy, but it should consumes it less and less.

Of the range of renewable energy options available, biomass offers the greatest potential both

---

[1] The actual index of residual life of the verified reserves of oil (39 years) and natural gas (62 years) achieves respectively 86 and 126 years, only the conventional technically recoverable resources are taken in consideration; 129 and 211 years, if the unconventional technically recoverable resources are considered.

in the short-term and overall as it offers the widest possible range of energy products and provides significant direct and indirect benefits (Berndes, Hoogwijk and van der Broek, 2003). These include:
- contribution to climate mitigation strategies by replacing fossil fuels;
- improvement of energy security;
- improvement of indigenous energy supply (reduction of energy imports);
- maintenance of agricultural and forestry economies with socio-economic benefits (employment opportunities);
- protection of biodiversity;
- potential for innovative scientific and technological developments.

The term bioenergy means that kind of energy derived from biomass. All combustible bio- logical material is a form of bioenergy: this may be wood, whole-tree chips, bark, chips, energy forest, energy crops, liquors from pulp production, domestic waste and industrial waste. In processed form, it may be an energy carrier such as, for example, electricity, heat, pellets, ethanol and biogas. The varied nature of biomass and the many possible routes for converting biomass resources to useful energy make this topic a complex subject.

It also becomes a matter of introducing and adding flexible energy technologies and design- ing integrated energy system solutions. Not only technological changes are required in order to generate further sustainable development: energy saving on the demand side, efficiency improvement in the energy production, replacement of fossil fuels by various sources of renewable energy, private-public cooperation, and so on (Lund, 2006).

1.1 Individual Motivations

Complex processes in individual life generate the **personal growth** that is colesely connected with assertiveness, self-esteem, creativity, emotional intelligence, problem solving, relationships, and communication. Energy itself is a **human right for self-improvement**: it increases the personal *capabilities* in the holistic sense of the Indian economist Amatya Sen (1985). Citizens have the right to expect energy in order to be able to develop their potential **individual capabilities**, which are reflexive and need sustainability (Donolo, 2001). In that sense, the citizens become crucial to a radical evolution of the whole society and the economy.

Energy can guarantee:

- **real freedoms** in the assessment of a person's advantage;

- **ability to transform resources** into valuable activities;

- **distribution of opportunities** within society.

Human development has multidimensional aspects and is based on the freedom of choice, on individual heterogeneity, and therefore also on energy access. The need for high-level capabilities in complex systems integration, project management, information technology, and advanced manufacturing techniques make energy solutions a natural fit.

Going in deeper in Nussbaum's specifications of the *ten capabilities*[2], it is clear how the energy plays a role in each of them: this research would emphasize particularly the political and material control over the environment (Naussbaum, 2000).

In the new era, the right to *a full life* is the **right to access**, which becomes the most important property value. People are able to generate their own energy, just as they create their own information and, with information, they can share it with millions of others. The same design principles and smart technologies that made possible the Internet, and the vast, decentralized global communication networks, could be used to reconfigure the world's power grids so that people can begin to **share energy peer-to-peer**, thus creating a **new, decentralized form of energy use**. People can control their own energy by organizing collectively to gain a modicum of energy independence (Rifkin, 2009).

The impossibility of accessing to energy is a key factor in perpetuating poverty around the world. Conversely, having access to energy means more economic and personal opportunities. **Electricity frees humans** from day-to-day survival tasks and increases their capabilities. **Distributed generation energy nets** connect communities and hold great promise for helping to lift billions of people out of poverty. If all the individuals and communities in the world become the producers of their own energy, the result would be a dramatic shift in the configuration of power. Local peoples would be less subject to the will of far-off centres of power. Communities would be able to **produce goods and services locally**. In that sense the energy is not only a mean but also a goal. The aim is to produce energy with quantitative (enough for equal personal development) and qualitative (safe and clean energy) characteristics.

1.2 Regional Motivations

Individual motivations allow to understand also the regional ones: the distributed generations of energy hold communities to become the producers of their own energy. Local areas would be less subject to the will of far-off centres of power. With distributed economies, a selective share of production is distributed to regions where a diverse range of activities are organised in the form of **small-scale flexible units** that are **emphatically connected** with each other and prioritise quality in their production.

Distributed economies offer a new strategy for designing road maps for regional development

---

2   The ten capabilities Nussbaum argues should be supported by all democracies are: life, bodily health, bodily integrity, senses, emotions, practical reason, affiliation, other species, play and control over one's environment.

and for the exploitation of the large wealth of gathered knowledge and innovation. Organizing regional activities allows mainly to have higher ownership and consequently to gain more power in directing these systems in ways that add quality to their lives. A big advantage of distributed economies is that it enables entities within the network to work much more with regional and local natural resources, finances, human capital, knowledge, technology, and so on. It also makes the entities more flexible to respond to the local market needs and thus generating a bigger innovation drive. By doing this, they become a better reflection of their social environment and in that way they can **improve quality of life**. Typically, socio-economic implications are measured in terms of economic indexes, such as employment and monetary gains, but in effect the analysis relates to a number of aspects, which include social, cultural, institutional, and environmental issues.

The relations in distributed economies are much more complex than those in a centralised economy: this feature makes the **whole economy more stable**, because leaf nodes no longer rely on just one central node.

Bioenergy production is interesting to take a driving role in the development of distributed economies. Thus, the energy conversion units are situated close to energy consumers, and large units can be substituted by smaller ones. **A distributed energy system is an efficient, reliable and environmentally friendly alternative** to the traditional energy system. The increased use of bioenergy, which exhibits both a broad geographical distribution, and diversity of feedstock, could secure long-term access to energy supplies at relatively constant costs.

To create **dynamically self-organizing business environments**, the deployment of the vast amount of globally available knowledge is necessary for the formation of regionally adapted strategies (Johasson et al., 2004). Similar, or better complementary, schemes can be brought together into networks to provide the advantage of scale without the drawbacks of inflexibility. Dynamic self-organization is vital for survival in a future of increasing complexity.

1.3 National Motivations

In addition to its local benefits, bioenergy production has **macroeconomic advantages** for countries: security of supply and an improved balance of trade for fuel-importing countries. The rapid increase in human population over the course of the XX century raises concern about whether Earth is experiencing overpopulation; the United Nations estimated that in 2011 the population will be 7,000,000,000[3]. That means also an increase in the usage of resources and energy, linked to social and environmental problems.

3   Population Division of the Department of Economic and Social Affairs of the United Nations Secretariat, June 2009.

The growing dependence of the European Union (EU) on imported oil has influenced several legislative initiatives (directives) intended to facilitate the development of biofuel markets. The greatest energy demand is actually met in two ways: by **increasing the exploitation of the planet's resources** and by **conducting a more intensive search for energy efficiency**. These solutions, however, create a vicious cycle that does not guarantee long-term sustainable development. A **holistich approach** to the sector can define **wide and continuously tuned solutions** to best suit dynamic and complex systems.

The energy sector is crucial for the development because it has economic, political, and social consequenses. The economical aspects are related to the competitiveness of a country and to the cost of raw materials, which take on a supranational role and create worldwide political and economic scenarios. In that way the economic aspects are strictly related to political aspects. The countries are (inter)dependent one to the other and those relations can change by the decision of few people, but can have consequences on all the population. The situation becomes worse when the resources come from non-renewable sources or sources that generate high environmental risks. Social aspects are related to those risks, the safety and the quality of life of the people.

From a macroeconomic perspective, bioenergy contributes to the important elements of a country's development:

- **economic growth** through business expansion (earnings) or employment;

- **import substitution** (direct and indirect economic effects on GDP);

- **efficiency improvement**;

- **security** of energy supply and **diversification** (Faaij, 2006).

These elements are strictly connected with the social ones, like the job creation, the increasing of personal capabilities and the wellness of people that live in a country.

Furthermore, bioenergy generation has many environmental benefits for an entire nation, such as the **reduction of Green House Gases** (GHG) and other pollutants, the **reduction of wasted biomass**, the **increasing of biodiversity**, and so on.

The emergence of small bioenergy units may ultimately radically change the way that the ever expanding electricity demand is met, especially in places currently poorly served by the traditional power system.

OVERALL STATUS OF THE ENERGY SECTOR IN EU

Energy is vital to modern industrialised society. Energy access, health impacts, energy security and climate change are all major challenges for achieving sustainable development, which are directly linked to energy systems (Geller, 2003). The **availability of affordable and reliable energy** allows many people to experience unparalleled comfort, mobility, and productivity (World Energy Assessment, 2000).

The key strategies and technologies identified in literature as the foundations for sustainable energy systems include enhancing energy efficiency, expanding renewable energy, improving fossil fuel technologies and advancing novel energy technologies. A major challenge confronting the EU and its Member States is **how to expand bioenergy** use to meet targets, policy goals, and international commitments on renewable energy, climate mitigation, energy security, and sustainable development.

In order to accomplish the EU's goal on renewable energy and GHG emissions the use of bioenergy has to expand[1]. Bioenergy improves energy security, combats climate change, and contributes to sustainable development. The potential for bioenergy in the EU is considerable, however, there are just a few Member States with sizeable bioenergy systems, and Sweden is an exception.

There is a range of supportive policy measures relevant for renewable energy in the EU. These policy measures are spread across several domanis: energy, agriculture and climate policy fields. For the past five years, the construction of a comprehensive EU energy policy based on mutually complementing pillars of competitiveness, sustainability and security of supply has significantly accelerated (*figure 2.0.1*).

The EU's energy production has been declining over the last five years, where the largest share being represented by nuclear (30%), followed by solid fuels (22%), gas (20%), oil (14%) and renewable resources (14%), although the contribution of the latter is expected to increase significantly in the future. In any case, the current fuel mix varies widely in the EU Member States; to a certain degree it depends on the domestic resource/production pattern.

The first step in this process was taken in 2005: the **Biomass Action Plan** (BAP), having as goals to accelerate the development of bioenergy, and to identify potentials and targets for biomass

---

[1] There are several EU policies on renewable energy with targets and timetables, including the achievement of a 22% share of electricity from renewable energy by 2010, and the expantion of the renewable energy share to 20% until 2020. There are also EU commitments in the Kyoto Protocol to reduce greenhouse gas emissions to 8% under 1990 levels by 2008-2012.

December 2005 — BIOMASS ACTION PLAN
potentials and targets for biomass resources

March 2006 — 1° GREEN PAPER
European strategy for sustainable, competitive, secure energy

October 2006 — ENERGY EFFICIENCY ACTION PLAN
coherent framework of legislations, politics and measures to achive the - 20% energy efficency goal

March 2007 — ENERGY ACTION PLAN
new legislation and proposals

January 2008 — 20-20-20 BY 2020 PACKAGE
reduction of GHG emissions and energy consumption, increasing of renewable energy resources

March 2008 — STRATEGIC ENERGY TECHNOLOGY PLAN
priorities for future energy technologies

July 2009 — ENERGY MARKET LIBERALISATION
"Third package" measures

November 2010 — 2° ENERGY 2020
European strategy for sustainable, competitive, secure energy

figure 2.0.I: main EU energy measures (2005-2010).

resources in terms of wood from forests, organic waste, wood industry residues, agricultural and food processing manure, and energy crops from agriculture. The BAP's objective is to double the current biomass contribution. The EU policies on biomass should be fixed on these concepts:

- systemic approach;

- subsidiarity and flexibility among Members States;

- competitive costs;

- competitive biomass uses in different sectors;

- review of legislation in Members States;

- development of National Action Plans for the biomass.

Furthermore, the European Environmental Agency states that reaching the targets for bioenergy as indicated in the BAP can be compatible (under proper management) with protecting and maintaining biodiversity, soil and water resources.

In 2006, the Commission approved the **Green Paper - A European Strategy for Sustainable, Competitive and Secure Energy**, which was revised in a second edition in November 2010: the **Energy 2020 - A European Strategy for Sustainable, Competitive and Secure Energy**.

Next, in October 2006, the Commission adopted a comprehensive **Energy Efficiency Action Plan** to create a coherent framework of legislation, policies, and measures for achieving the 20% energy efficiency objective compared to what was expected to happen under a business as usual scenario. However, the quality of the National Energy Efficiency Action Plans, developed by Member States since 2008, is disappointing, leaving vast potential untapped. The move towards renewable energy use and greater energy efficiency in transport is going too slowly.

In January 2007, the Commission issued the first EU **Energy Action Plan**, which was endorsed by the European Council in March 2007. At present, most of the Energy Action Plan measures have been largely executed through new legislation and ongoing proposals that will soon be agreed upon. In September 2007, in order to complete the integration of the EU gas and electricity market, the Commission proposed further **energy market liberalisation** measures (third package). The Council and the Parliament agreed on this legislation in July 2009.

The **Strategic Energy Technology Plan**, presented by the Commission in November 2007 and agreed upon in March 2008, introduced the priorities for future energy technologies.

In January 2008, the Commission proposed the Energy and Climate package with "**20-20-20 by 2020**" goals: a reduction in GHG emissions up to 20% below the 1990 levels, 30% in the context of a global agreement on climate; a 20% share of renewables in the final energy consumption; a reduction in primary energy use up to 20% below the baseline projection for 2020; goals that were subsequently translated into legally-binding frameworks for GHG emissions and renewable energy. The focus of this research is the second point: **design a framework to encourage and to represent the production of energy from renewable resources**.

In 2008, an action plan was put forward in the Second Strategic Energy Review. It emphasized the importance of infrastructure links needed to strengthen energy security and solidarity between Member States, as well as introduced the perspective of a low carbon economy to be achieved by 2050, which will necessitate a major shift towards low carbon energy technologies.

Additionally, the policy goals are related to many topics, including liquid biofuels for transport, electricity and heat from renewable energy, GHG emissions, and energy crops. The types of policy measures range from the Kyoto Protocol to the Common Agricultural Policy and to the EU Directive on the Promotion of the Biofuels for Transport.

Many of the regions assisted by the structural and cohesion funds have high potential to pursue economic growth and employment creation or stabilisation through biomass. This is particularly true for rural regions in Central and Eastern Europe. Low labour costs and high resources availability can give these regions a comparative advantage in the production of biomass. Supporting the development of renewable and alternative energy sources such as the production of biomass is therefore an important objective for the structural and cohesion funds. These funds may support the retraining of farmers; the provision of equipment for biomass producers; investment in facilities to produce biofuels and other materials; and fuel switching to biomass by electricity and district heat producers.

The EU energy and climate goals have been incorporated into the Energy 2020 Strategy for smart, sustainable and inclusive growth, adopted by the European Council in November 2010, and into its flagship initiative *"Resource Efficient Europe"*. To achieve the many stipulated targets will require the development of effective policy measures that can stimulate innovation processes and engage key stakeholders.

2.1 Criticalities

Bioenergy has an obvious opportunity to play an important role in the energy systems of the future, but it will not suffice for all the various applications that may arise in the future. Unfortunately, the current projections show that the EU is lagging behind the widely accepted BAP targets, which are based on the use of existing technologies and systems, because the Members States need to intensify the effort to expand bioenergy. Based on the available literatures, case studies, and industry interactions, this thesis work identifies the key barriers to the sustainable energy development (*table 2.1.I*). **EU is facing three main barriers in the development of green energy generation** (McCormick, 2007):

- **Economic conditions**: renewable energy should compete with fossil fuels and nuclear power,

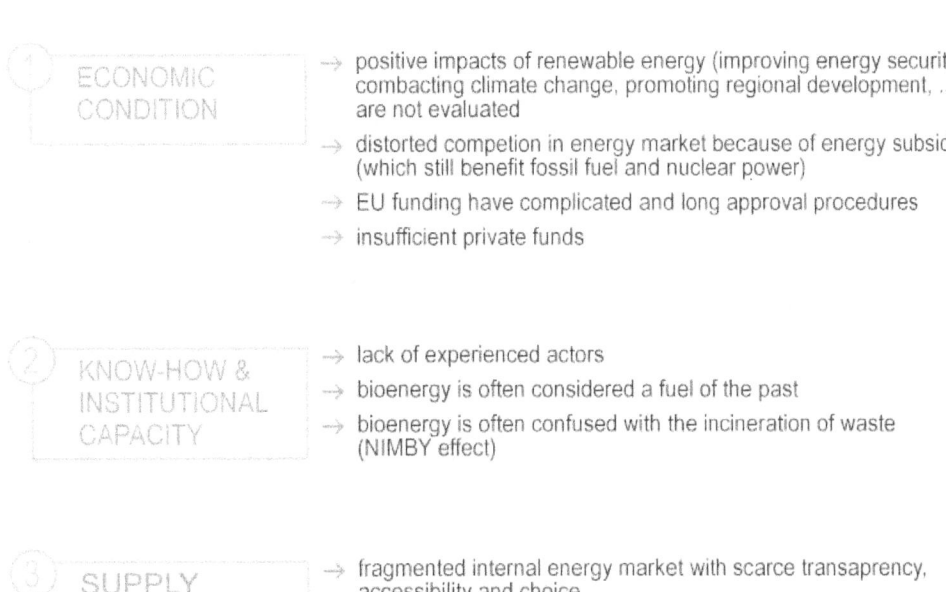

table 2.1.I: non-technical criticalities in EU green energy generation development.

which have received and continue to benefit from energy subsidies and externalized costs. Renewable energy often produces positive impacts that are not compensated by energy markets, including improving energy security, fighting climate change, and promoting regional development (Boyle, 2004). Those benefits are rarely recognized with supportive policy measures (incentives and disincentives) and by investors. There is a distorted competition on energy markets linked primarily to energy subsidies, which still privilege fossil fuels and nuclear power. Economic conditions represent both a broad context and specific issues for investors. Facing economic uncertainties and market risks on one side and the higher costs for the provision of energy from biomass compared to the energy costs from fossil fuel on the other side, a broad and long lasting market introduction programme is required to build up a bioenergy market. The EU gets funds to implement bioenergy projects, but sometimes the funding conditions can change within the duration of the funding or even the application period. Long and complicated approval procedures can lead to a significant time lag between an application for funding and the final decision. Such time delays can make projects even more expensive. Also the private financings through banks or other financing institutions are not so easy to obtain due to given technical and non-technical uncertainties (in the early stage of development).

- **Know-how and institutional capacity**: a combination of know-how and institutional capacity is necessary to shift from unrealized potential to success. The lack of understanding of the bioenergy industry by the financial sector may prove to be a barrier, as well as the absence of experienced maintenance staff. The required know-how may be developed in existing actors through learning processes and by the introduction of new actors. For example, in bioenergy expansion, the energy crops play an important role, but the farmers often have minimal experience regarding them. Uncertainty based on a lack of experience discourages many farmers from investing in energy crops. Even when farmers are convinced by the long-term viability of energy crops, they are often too concerned by the short-term effects on their work practices and economic flows. Furthermore, the lack of experience is also noticeable within the institution that should give the authorizations. Learning processes and altering the perceptions of both public and politicians about renewable energy are often required to build up the legitimacy for that kind of production. Bioenergy is frequently considered a fuel of the past rather than a fuel of the present and the future (Hall & Scrase, 1998) and is often confused with the incineration of waste[2]. Bioenergy plants suffer the NIMBY (Not In My Back Yard) effect, because they need to be built in urban or industrialised area. For that reason the social acceptance and comptence of the plant owner and operator within a community are often an important factors for the public acceptance of a project. Cooperation between public and private actors bundles financial resources, know-how, and expertise.

- **Supply chain coordination**: the complex systems that involve many actors require functioning and organized supply chains that overcome the "chicken and egg" problems[3]. Bioenergy systems require functioning and organized supply chains that meet the needs of all relevant actors (energy companies and biomass suppliers). Investing in biomass resources is generally possible if there are energy companies purchasing biomass. In addition, establishing conversion technologies is generally only possible if there are biomass suppliers providing materials. Technologies and systems are required for harvesting, refining, and transporting biomass. Current bioenergy activities are integrated with other industries by using the other industries's waste as fuel; the using of existing skills or structures, such as machines and forest roads, gets the business secure and cheap. Contracts with partners and subcontractors need to be signed to regulate the cooperation of the different participants to carry out a successful project during the various stages of the overall project lifetime. The internal energy market is still fragmented and has not achieved its potential for transparency, accessibility, and choice. Companies have grown beyond national borders, but their development is still hampered by a host of different national rules and practices.

The main consideration is that non-technical issues dominate the key barriers that can delay

---

[2] Different definitions and legislation for waste across the Members States can present problems. When feedstock is categorized as waste this often means more stringent legislation (and different reactions from the public and the politicians).

[3] It highlights the challenge of investing in biomass resources at the same time as establishing conversion technologies. Neither can proceed without the other but it is difficult to draw up contracts that are acceptable to both energy companies and biomass suppliers.

the planning and realization of economically feasible biomass energy projects. In general, more time is needed to implement a bioenergy system compared to an energy system based on fossil fuels (Rösch, Kaltshmitt, 1999). This is due to the significantly higher complexity of biomass energy projects involving a great number of different partners compared to a fossil-fired plant with proven technology and a clear procedure on how to construct and run such a plant.

A better understanding of critical factors, their interactions, and importance for the implementation of bioenergy systems is useful for any actors interested in the development of bioenergy and gives an analytical perspective. The purpose of analysing key barriers is to identify the challenges of primary importance that Members States need to compare in order to expand bioenergy. This research will take into account all these problems to advance a new model of green energy productions to develop Local Economic Development (LED) using the SD methodology.

2.2 Future potentialities

Bioenergy is the only one of today's renewable alternatives that naturally gives carbon-based fuels, on which large parts of the energy technology are also based. It is also e relatively cheap energy source compared to other climate-friendly alternatives. In the future there will an increased demand and also availability of biomass (Berndes, Hansson & Wirsenius, 2008), even if it represents only one part of the sustainable energy strategy for EU.

The recent commitments made by the EU show growing political leadership on this issue: the established targets of the EU represent a roadmap for the expansion of renewable energy.

The EU fixes some specific goals to be reached in the next decades in order to solve the actual potentialities of success (*table 2.2.I*):

- the **energy efficiency:** there is a significant potential for reducing consumption, especially in energy-intensive sectors. Mobilising the public opinion, decision makers and market operators and setting minimum energy efficiency standards and rules on labelling for products, services and infrastructures are only the beginning. Effective compliance monitoring, adequate market surveillance, widespread usage of energy services and audits, as well as material efficiency and recycling are all musts.

- the **greater diversity of input fuels**: the reliability of the EU's generation capacity will also be strengthened by a more diverse generation mix. Diversity may be a business and a policy aims at the same time since it avoids risk exposure, provides flexibility to accommodate variations in demands, and helps to reduce heavy dependency on one type of fuel and/or on one technology. Diversity may thus foster reliability and greater interdependence. It is crucial for energy security and should be considered when assessing the adequacy of power generation infrastructure in the EU.

- the **reduction of import dependency** on one supplier or in state of monopoly (more than 80%): the energy savings and diversification improvements with more renewables will make the EU less vulnerable to the effects of volatile import price developments. Managing the import dependency is thus an issue that has economic, political and security dimensions.

- the **cleanliness of new energy generation**: the low-carbon power generation should also be cost-effective and reliable, with a further move towards small-size and decentralized units. Those aspects are strictly related to research, technology development, and demonstration.

- the **flexibility in infrastructures** to withstand possible supply shock and to make possible the diversification of suppliers to other directions: today's grid will struggle to absorb the volumes of renewable power which the 2020 targets entail. The EU is still lacking the grid infrastructure which will enable renewables to develop and compete on an equal footing with traditional sources.

- the **smart energy network**: the integration of Distributed Energy Resources within the European electricity network is already started, supported by the recent strengthening of the EU energy policy, and significant market and technology evolutions. New communication solutions, amongst others wireless technologies coupled with smart sensors and smart meters will pave the way to improved network automation, facilitating the use of Distributed Generation and Demand Response. To facilitate the transition to a more sustainable energy system, a wide-ranging Research & Development (R&D) effort is required on EU electricity and networks. Research aims at the effective integration of biomass installations into electricity grids and feeding biogas and synthetic natural gas into the natural gas grid.

- the **simplification of the procedures to obtain authorization** for new plants: the procedures to obtain permission from the government to run biomass energy plants depending on the size, process technologies, and feedstock should be improved substantially and become simple, less time consuming, and less costly, as well as more transparent. Communities at local, regional and national level will engage more constructively in facilitating projects of common interest if they also bring concrete and short term benefits.

The supportive policy measures established by the EU indicate a political willingness to bring about change in current energy systems. The increase use of bioenergy, in particular, remains at the forefront of the EU strategy to respond to climate change and improve energy security. A key political challenge is maintaining the support for biomass while, at the same time, harmonising taxes and other policy instruments across countries and sectors. A key policy challenge is maintaining and expanding the use of biomass and bring about equally cost-efficient and rational uses for other energy purposes in the longer term.

The EU is the level at which the energy policy should be developed. Decisions on the energy

table 2.2.I: EU goals.

policy taken by one Member State inevitably have an impact on other Member States. The optimum energy mix, including the swift development of renewables, needs at least a continental market. Energy is the market sector where the greatest economic efficiencies can be made on a pan-European scale. Fragmented markets not only undermine the security of supply, they also limit the benefits which the energy market competition can bring. The new EU energy strategy will require significant efforts in technical innovation and investment. It will foster a dynamic and competitive market and will lead to a major strengthening of institutional arrangements to monitor and guide these developments. It will improve the security and the sustainability of energy systems, grid management, and energy market regulation. It will include extensive efforts to inform and empower domestic and business consumers, to involve them in the switch to a sustainable energy future, for example by saving energy, reducing wastage and switching to low-carbon technologies and fuels. The new strategy will take the first steps to prepare the EU for the greater challenges. Above all, it will ensure better leadership and coordination at the European level, both for internal action and in relations with external partners.

PART II

RESEARCH METHODOLOGY

SYSTEMIC DESIGN THEORY

3.1 What

3.1.a History: from generative science to Systemic Design

On the basis that living systems continually draw upon external sources of energy and maintain a **stable state of low entropy**, as the physicist **Erwin Schrödinger** asserted in **1946**, some of the next theories on industrial processes applied that concept also on **artificial systems**. In that way the field almost exclusively populated by engineers, has been diluted with the increasing involvement of different disciplines including design, urban planning, public policy, business management and environmental sciences (Chertow, Ashton, & Kuppali, 2004).

Generative science is the interdisciplinary and multidisciplinary science that explores the natural world and its complex behaviours as a generative process. Generative science shows how finite parameters in the natural phenomena interact with each other to generate infinite behaviours. This science explores the natural phenomena at several levels including physical, biological and social ones (*figure 3.1.I*). Generative science originates from the monadistic philosophy of **Gottfried Leibniz**, at the **end of XVII century**, in which monads are the ultimate elements of the universe: "*substantial forms of being*". This was further developed by the neural model of **Walter Pitts** and **Warren McCulloch**[1], in the **early XX century**. The development of computers laid a technical source for the growth of the generative sciences. However, the cornerstones of the generative sciences came from the work on cellular automaton theory by **John Von Neumann**[2], in the **1940s**. The generative sciences were further unified by the cybernetic theories of **Norbert Wiener**[3] and the information theory of **Claude E. Shannon** and **Warren**

---

1   Walter Pitts was a logician who worked in the field of cognitive psychology and Warren Sturgis McCulloch was an American neurophysiologist and cybernetician. They wrote a seminar paper entitled "A Logical Calculus of Ideas Immanent in Nervous Activity" (1943), that proposed the first mathematical model of a neural network. The unit of this model, a simple formalized neuron, is still the standard of reference in the field of neural networks.

2   A cellular automaton is a discrete model studied in computability theory, mathematics, physics, complexity science, theoretical biology and microstructure modelling. It consists of a regular grid of cells, each in one of a finite number of states, so it can be in any finite number of dimensions.

3   Cybernetics is the interdisciplinary study of the structure of regulatory systems; it is equally applicable to physical and social systems. Cybernetics is pre-eminent when the system under scrutiny is involved in a closed signal loop, where action by the system in an environment causes some change in the environment and that change is manifest to the system via information, or feedback, that causes the system to adapt to new conditions: the system changes its behaviour.

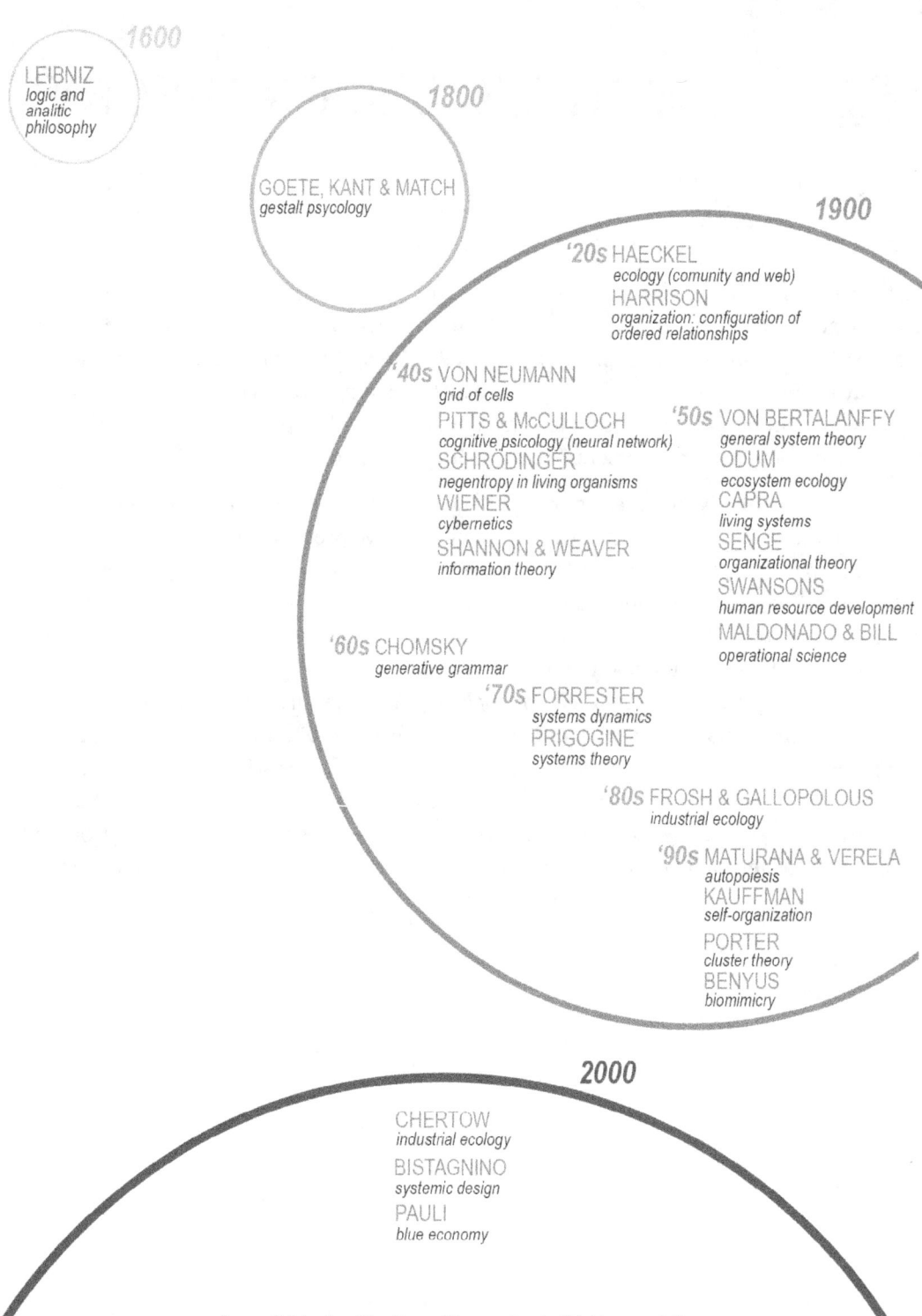

figure 3.1.l: simplification of the systemic thinking evolution.

Weaver[4] in **1948**. They built the idea of unifying the physical, biological and social sciences into a holistic discipline of **Generative Philosophy** under the rubric of **General Systems Theory** (GST) by **Ludwig von Bertalanffy**[5], that in **1954** established the **Society for the Advancement of GST** with **Anatol Rapoport, Ralph W. Gerard, Kenneth Boulding**. He stated that "*a system is a set of unities with relationship among them*", so it is evident the relational aspects among the several parts and the global essence of the whole system.

Contemporary ideas from systems theory have grown within diversified areas, exemplified by the **ecosystem ecology** by **Eugene Odum**, the **living systems** by **Fritjof Capra**, the **organizational theory** by **Peter Senge**, the interdisciplinary study about areas like **Human Resource Development** by **Richard A. Swanson** and **insights from educators** such as **Debora Hammond**. As a transdisciplinary, interdisciplinary and multi-perspective domain, it brings together principles and concepts from ontology, philosophy of science, physics, computer science, biology and engineering as well as geography, sociology, political science, psychotherapy and economics among others.

Generative science had an important influence by the development of the **cognitive sciences** through the **theory of generative grammar** by **Noam Chomsky**[6], in the late **1950s**.

In **1977 Ilya Prigogine** received the Nobel Prize for his works on **self-organization**, conciliating important systems theory concepts with system thermodynamics. It has advanced in the field of the **autopoiesis** by **Humberto Maturana** and **Francisco J. Varela**[7] and in **self-organization** by the works of **Stuart Kauffman**[8], in **90s**. Generative scientists are working towards further developments and new frontiers. Latest and emerging directions in these sciences in-

---

4   Information theory is a branch of applied mathematics and electrical engineering involving the quantification of information. A key measure of information in the theory is known as entropy, which is usually expressed by the average number of bits needed for storage or communication. Intuitively, entropy quantifies the uncertainty involved when encountering a random variable.

5   Bertalanffy originated the term General System Theory to identify the transdisciplinary study of systems in general, with the goal of elucidating principles that can be applied to all types of systems in all fields of research. It is a specialization of systems thinking and a generalization of systems science.

6   A generative grammar of a language attempts to give a set of rules that will correctly predict which combinations of words will form grammatical sentences. He has argued that many of the properties of a generative grammar arise from an "innate" universal grammar.

7   They were the first to define and employ the concet of autopoiesis; aside from making important contributions to the field of evolution and constructivist epistemology; an epistemology built upon empirical findings of neurobiology.

8   He argues about the complexity of biological systems, the evolution and the self-organization of organisms.

clude the computer simulations[9] of complex social process and artificial life (i.e. Boids[10]).

Complexity models of living systems address also productive models with their organizations and management, where the relationships between parts are more important than the parts themselves. Treating organizations as **complex adaptive systems** allows a new and more productive management model to emerge in economical, social and environmental benefits (Pisek & Wilson, 2001). In that field, **Cluster Theory** (CT) evolved in more environmental sensible theories, like **Industrial Ecology** (IE) and the **Industrial Symbiosis** (IS), and finally in more methodological theory: SD.

3.1.b Principles

Cluster Theory (figure 3.1.II)

Clusters have long been part of the economic landscape, with geographic concentrations of trades and companies dating back for centuries. The intellectual antecedents of clusters date back at least to Marshall, who included a chapter on the externalities of specialized industrial locations in his *Principles of Economics* (1920).

Knowledge about CT has advanced in the field of competitiveness, economic development and environmental policy by Michael E. Porter[11] (1990).

Cluster is a geographically proximate group of interconnected companies and associated institutions in a particular field, linked by analogies and complementarities (reducing externalities). The industries can share resources (i.e. knowledge management), so it is called horizontal cluster; or the suppliers, so it is called vertical cluster.

Even if old reasons for clustering have diminished in importance with globalization, new influences of clusters on competition have taken on growing importance in an increasingly complex, knowledge-based, and dynamic economy, because distant rivals cannot match (Porter, 1998). Clusters support the economic development by building on local differences, seeking an endogenous growth of regional economies, reinforcing the assets already present in the local economies.

Clusters create competitive advantage in three ways:

- by increasing the productivity of companies involved in the cluster;

---

9   It is a computer program that attempts to simulate an abstract model of a particular system. The scale of event being simulated by computer simulations has far exceeded anything possible using the traditional paper and pencil.

10   It is an artificial life program, developed by Craig Reynolds in 1986, which simulates the flocking behaviour of birds (emergent behavior). The complexity of Boids arises from the interaction of individual agents adhering to a set of simple rules (separation, alignment, cohesion).

11   It is also known as Porterian Cluster, Competitive Cluster, Business Cluster.

- by driving the direction of innovation, which underpins future productivity growth;

- by stimulating new businesses, which expands and strengthens the cluster.

However, the competition among co-located industries could be avoided if companies with similar products, but different niches, can benefit from each others' competence and shared network of suppliers and service companies (Wolf, 2007).

Cluster boundaries rarely conform to standard industrial classification systems, which fail to capture many important actors in competition and linkages across industries, because parts of a cluster often are put into different traditional industrial or service categories. Those boundaries continually evolve as new firms and industries emerge, established industries shrink or decline, and local institutions develop and change. For CT the price is the sole competitive variable, and firms hold down wages to compete in local and foreign markets.

However, CT is seldom focused on environmental issue although environmental policies can give to the involved companies a competitive edge.CT and IE have a lot in common, but they rarely interact, with the exception of Deutz and Gibbs, who use an empirical focus on eco-industrial development in USA to postulate that IE can be considered a distinct cluster concept (Deutz & Gibbs, 2008).

CLUSTER THEORY (CT)

1990: Porter

1. industries in the same sector

2. links among industries based on commonalities & infrastructures

3. geographical concentration

4. **evolution of boundaries**

figure 3.1.II: CT main principles.

Industrial Ecology *(figure 3.1.III)*

The IE emerged in 1989 after an article by Frosch and Gallopoulos in Scientific American; but the foregoing idea was from the 50s and its early stages of development was in the mid-70s (Desrochers, 2002). The IE is a model to increase efficiency and to reduce resource consumption and disposal, mimic the Nature. It symbiotically links industries so that environmentally conscious practices can also be profitable. To do this, IE uses principles of biological ecosystems to optimize the flows and transformations of materials and energy within and across the boundaries of industrial systems in holistic view[12]. The principles of natural systems provide a scientific framework for understanding and creating more sustainable urban systems. Natural systems are cyclical and relatively efficient. Matter and energy are neither created nor destroyed. Waste materials from one organism serve as food for other organisms; energy flows through each of the trophic level in the food chain (Cosgriff, Dunn, & Stainemann, 1998).

IE models urban systems on natural systems to increase efficiency and to minimize demands on environmental sources and sinks (Socolow *et al.*, 1994). IE takes a systemic view of production and consumption activities, rather than focusing specifically on waste reduction activities on an industry-by-industry basis (Allenby, 1994).

The theory of IE offers a set of guidelines to target the business community (Tibbs, 1993; Ehrenfeld, 1997):

1- **Balancing industrial input and output** to the constraints of natural systems: identifying ways that industry can safely interface with nature, in terms of location, intensity and timing; and developing indicators for real-time monitoring. In natural systems, a species can evolve to fill a particular form or function, or a niche. Similarly, in industrial systems, niche industries can evolve to utilize or transform industrial waste into useful inputs. This systemic view seeks to better integrate industrial processes with each other, and with the natural system.

2- **Dematerialization of industrial output**: striving to decrease materials and energy intensity in industrial production.

3- **Improving the efficiency of industrial processes**: re-designing production processes for a maximum conservation of resources. Biological processes replace chemical and technical processes, because learning from biosphere can provide sustainable technical processes.

4- **Symbiotic links among industries**: promoting both environmental quality and economic development, and thus sustainable communities in several ways. Firstly, waste products from one industry can provide input materials for another industry, which can reduce the

---

[12] This definition of Industrial Ecology was developed by the Industrial Ecology Study Group at the Center for Sustainable Technology, Georgia Institute of Technology, 1997.

INDUSTRIAL ECOLOGY

1989: Frosh & Gallopoulos

1. balancing input/output to the contraints of natural system

2. dematerialization of industrial output

3. improving the efficiency of industrial processes

4. symbiotic links among industries
   cooperation

5. creating new structures

6. industrial eco system

- **close loop**
  cycle

- **holistic view**

- **flows of** — *material* / *energy*

- **benefits** — *social* / *economical* / *environmental*

figure 3.1.III: IE main principles.

receiving industry's cost at resource inputs. Economic theory suggests that as non-renewable resources become scarcer, their prices will rise; accordingly, reduced reliance on non-renewable resources can reduce costs of production. This can also promote environmental quality because it reduces the use of raw materials. Secondly, IE has the potential to reduce waste disposal costs because, now, waste products serve as input for other industrial processes. This can reduce pollution because industrial by-products are reused, rather than discharged directly. Thirdly, IE can increase industry profits because materials previously regarded as waste now have an economic value. Niche industries can evolve to fill in the gap between other industries to utilize and sell waste materials in an exchange of energy and material flows.

5- **Creating new structures**: coordinating actions, communicative linkages and information exchanges. IE creates jobs with a larger and more varied employment base (PEZZOLI, 1997). In addition, reductions of industrial emissions could decrease the need to separate industrial and residential land uses, allowing workers to reside closer to employment centres.

6- **Industrial ecosystem = eco industrial park**: working areas where IE is working.

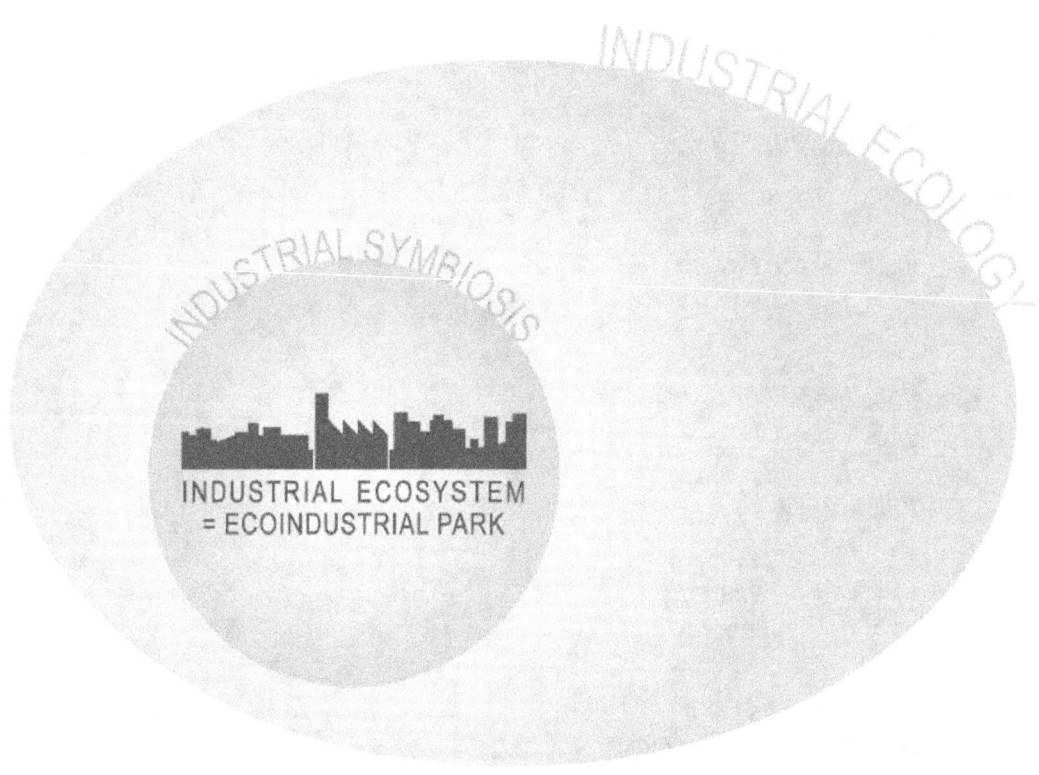

figure 3.1.IV: relations among IE, IS and inductrial ecosystem.

Communities of businesses cooperate with each other and with the local communities to efficiently share resources, leading to economic and environmental quality gains and equitable enhancement of human resources. In an industrial ecosystem, the effluents of one industrial process serve as the input materials for other industrial processes, which parallels food webs in natural systems, in a very efficient way (*figure 3.1.IV*).

IE is not the only solution, but it is a particularly compelling one because it provides economic incentives to reduce the adverse environmental impacts of industrial and municipal production processes.

Although IE is a promising approach, it is not a panacea.

One concern surrounding IE is that industries will use it as an alternative to pollution prevention initiatives. If a profit stream results from a waste stream, this could provide incentives for increased resource extraction or waste production. In addition, industries' agreement to exchange by-products could perpetuate their reliance on toxic materials. One possible response is that IE is intended to reduce system-wide pollution through industry-level pollution management.

A second concern regards competitiveness: companies risk losing a critical supply or a marketplace if a plant closes down or changes it product mix, or if proprietary information become available to competitors (Lowe, 1997). Because the success of the system depends on the success of each participants, designer should consider the delicate nature of these arrangements and be sensitive to confidentiality issues.

A third concern, particularly for material flows, is the chemical integrity of the by-product being exchanged. Manufacturing processes often combine materials, which may reduce the likelihood of any manufacturing plant to be able to account for the exact chemistry of their waste stream. Caution must be exercised in instances where industrial by-products will be used, for example, as feedstock or fertilizer.

A final concern points to a need for designing to plan for integrate IE within a broader framework of community sustainability.

Industrial Symbiosis (figure 3.1.V)

The IS is a part of IE and is regarded to pursue IE's objectives within geographically coded areas. It demands resolute attention to the flow of materials and energy through local and regional economies.

IS traditionally separates entities in a collective approach to competitive advantage involving physical exchanges of materials, energy, water and by-products (Chertow, 2000).

Different industries collaborate among them for mutual economic and environmental benefit, even if partners should be independent ("across the fence"). Someone's waste is one's raw material, in a way that is economically and environmentally profitable. The IS is the development of industries in a system to reach improved performance. This is because exchanges enabled through collaborative synergistic connections have the potential to improve resource use efficiencies, thus contributing to the reduction of resource throughput and pollutant generation.

The symbiosis means that the system works in mutualism and commensalism. The synergies among industries is fundamental because "working together" (from the greek syn-ergo) the final outcome is greater than the sum of the parts. Actors are organized into networks with integrated material and energy flows. The keys are collaboration and synergistic possibilities of-

INDUSTRIAL SYMBIOSIS
2000: Chertow

1. **output > input**

2. **symbiotic links among industry**
   mutualism and commensalism

3. **self-organization**

4. **geographic proximity**

+ **flows of**
  - material
  - energy
  - water
  - by-products

+ **benefits**
  - economical
  - environmental
  - social

figure 3.1.V: IS main principles.

fered by geographic proximity (Chertow, 2007), so that is possible to share not only tangible resources, but also, for example, logistic infrastrutures and managerial or knowledge resources.

An important contribution of that theory is that IS is self-organized (Hardy and Graedel, 2002) driven by economic advantages offered by market dynamics and/or policy requirements.

Systemic Design (figure 3.1.VI)

The methodology of SD looks at making better use of material and energy flows in order to model our production and energy systems after nature (Bistagnino, 2009). Material and energy loops are open in order to decrease environmental impacts and resource depletion. Living systems are "open" in the sense that they continually draw upon external sources of energy and maintain a stable state of low entropy that is far from thermodynamic equilibrium (Shrodinger, 1943). Many industrial ecosystems have come about ad hoc for better business, while others have been facilitated through external actors. However, as these theories and ventures may be innovative for the industries, they are still no more than solving problems that arisen from environmental pressure and economical revisions.

Systemic theory is the study of how complex entities interact openly with their environments and evolve continually by acquiring new, "emergent" properties (Heylighen et al., 2000). Rather than reducing an entity to the properties of its parts or elements, systems theory focuses on the relationships between the parts that connect them into a whole. Complex systems are generally dynamic, nonlinear and capable of self-organization to sustain their existence. This approach is patterned after the self-organizing behaviour of living systems. This type of reasoning leads to the "Gaia hypothesis", which claims that the world is a single giant organism (Lovelock, 1988).

SD proceeds with constant awareness of related systems, boundary conditions, external effects and potential feedback. SD plans entities with inherent "resilience" by taking advantage of fundamental properties such diversity (existence of multiple forms and behaviours), efficiency (performance with modest resources consumption), adaptability (flexibility to change in response to new pressures) and cohesion (existence of unifying forces or linkages). (Fiksel, 2003)

To encourage systems design that incorporates sustainability thinking explicitly, it is useful to have clear principles. The theory of SD offers a scientific method that derives from the generative science and evolves from IE, symbiosis, ecosystem and CT. It can be summed up by its five basic principles (Bistagnino, 2009):

1- **Output>Input**: as in nature what is not used by a system becomes a raw material for the development and survival of someone/something else, in the production process the waste (output) of a system become an opportunity (input) for another one, creating new economic opportunities and new jobs.

2- **Relationship**: it is important to consider, more broadly, all the networks of components that made the food system, including materials (resources) and energy, which are used, captured and stored through different stages of the product life cycle. Understanding the pattern of materials and energy flow and investigating where it can be improved can allow us to find entry-points for designing more sustainable food system.

SYSTEMIC DESIGN
2009: Bistagnino

1. output > input

2. relationship

3. autopoiesis

4. act locally

5. man at the center of the project

flows of — material / resources / information

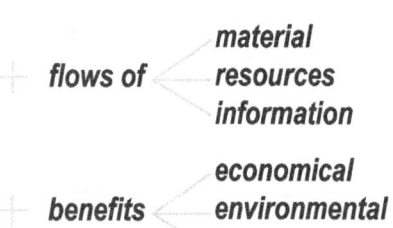

benefits — economical / environmental / social

figure 3.1.VI: SD

3- **Towards autopoiesis**: in nature self-maintaining systems sustain themselves by reproducing automatically, thus allowing them to define their own paths of action. In this way the system is naturally led to balance and preserve its independence. If also in the food system we would start in terms of autopoiesis, it could be possible to allocate efficiently and distribute equally the material and energy flow.

4- **Act locally**: as an eco-system is deeply influenced and shaped by its habitat, the same happens for any other type of system. Based on the opportunities provided by the local context, new opportunities can be created by reducing the problems of adaptability due to "general" solution and increasing people's participation.

5- **Man at the centre of the project**: the product has become fulcrum of a paradigm of values and actions, as the economical wellness, the quantity of currency resources, the wish of belonging to a social status, that shape negatively consumption choices. The systemic approach, instead, questions the present industrial setting and proposes a new paradigm where at the centre of each productive process there are social, cultural, ethical and biological values that are every man shares.

On an industrial level, the logical linear process and development affect the perception of reality, as they are based merely upon cause-effect relationships, which generate huge quantities of waste, starting with the manufacturing process until the product "end of life".

With such on-growing complexity, it is necessary to withdraw from the exclusive focus given on the product and its life cycle, moving toward achieving greater competence in the context of the complex relationships which spring from the production process.

Bring back within the total planning equation, the variable represented by those resources (generated as a result product or waste) would otherwise be unused.

The cultural and practical skill should be regained to outline and plan the flow of matter, running from one system to another, in an on-going metabolizing process which diminishes the ecological imprint and generates a remarkable economic flow. Currently rejects generated during the manufacturing process, are only a cost.

In order to go through a project with the SD approach, it is essential to start from the current state and to make a peculiar observation of all the aspects which are part of the system (input), what occurs inside it and what comes out from it (output).

The analysis of these inputs and outputs will have to be of two different ways:

- **quantitative** so as to know the quantities that we move around and we avail of;

- **qualitative** to know exactly what we have at our disposal.

These simple steps enables to have a clear idea of:

- the resources we need, their features and origins;

- rejects or processing waste, their specific qualities and their final destination;

- what occurs throughout the processes, comparing the specific differences of inputs and outputs.

The vision holistically embraces the whole process and makes one perceive the relationships interwoven within the analysed system. This shows how the mere focus on individual parts, oblivious of the links with its component parts, is utterly useless, and forms a dichotomy with the dynamism of the whole.

What explained so far aims to make one understand the actual cause of the malfunction in the current cultural and productive approach of the society, only geared towards achieving products and solving specific problems, ignoring the connections occurring among the constituent parts or the cause-effect relationships of the choices we make.

For an overall vision, it is necessary to arrange a graphic scheme, allowing to retrace both with eyes and mind, the flows of matter and energy into the system we have looked at until now.

From the analysis of the problems and the configuration of the starting scheme, the quality of the outputs are highlighted, so they can be turned into inputs for other productions or systems, looking for possible links with other territorial realities which, despite their great differences, can be integrated within the analysed production.

This process usually leads to an exponential growth of the productive capacity of a territory, as a result of which, it is able to produce new material goods, to offer new services to the citizens and to ultimately increase the number of jobs. Yet, contextual surveys show a demand actually exists but it is met by imports or external resources. By exploiting the resources of the territory, the development has a local dimension and new self-sustained realities are spawned, both in terms of energy, production and supply.

3.2 Why

Currently, the society faces situations by cause-effect phenomena, solves technical problems by study strategies "*per spot*", using a linear approach. The knowledge to be spread does not exclusively refer to the process of putting up a product where the only aesthetic features is the enhanced one, but it is the awareness of the system related to products and where they have been generated. On an industrial level, the logical linear process and development affect the perception of reality, as they are based merely upon cause-effect relationships, which generate huge quantities of waste, starting with the manufacturing process until the product "end of life".

With such an on-growing complexity, it is necessary to withdraw from the exclusive focus given on the product and its life cycle, moving toward achieving greater competence in the context of the complex relationships which spring from the production process.

A comprehensive systemic approach is essential for effective decision making with regard to global sustainability, since industrial, social, and ecological systems are closely linked. Despite the efforts to reduce un-sustainability, global resource consumption continues to grow. There is an urgent need for a better understanding of the dynamic, adaptive behaviour of complex systems and their resilience in the face of disruptions, recognizing that steady-state sustainability models are simplistic. A number of research groups are using dynamic modelling techniques, including bio-complexity, system dynamics, and thermodynamic analysis, to investigate the impacts on ecological and human systems of major shifts such as climate change and the associated policy and technology responses. These techniques can yield at least a partial understanding of dynamic system behaviour, enabling a more integrated approach to systems analysis, beneficial intervention, and improvement of resilience (Fiksel, 2006).

3.2.a Historical Background

The evolution of the concept that industrial processes can work like Nature shows how the SD theory adds some crucial elements to previous theories making them more effective.

The CT is far away from what is now SD, because it is totally focused on the geographical concentration of companies that work in the same sectors to share analogies and infrastructure, with no environmental consideration (*figure 3.2.1*). The theory is totally business oriented and the strategic decision to settle companies in the same area is taken to reduce costs (externalities) and to take advantages on the market by driving innovation. This is the result of a totally different paradigm in theories: the CT is mainly focus on products and SD is focus on human beings. The values generated by these two paradigms change the relational and evaluative priorities in social and productive contexts. The focus on products generates values intrinsically related to it, which does not consider the person and the environment, but only the economical profit. By using that kind of analysis becomes clear why CT does not deal with output>input concept and environmental benefits. Furthermore, the geographical concentration, that is one of the main issues of this theory, is not examined for the territorial valorisation, but just for cutting cost. In that sense, it is not so important whether Silicon Valley is in California or in another part of the world, while the fact that Apple, Cisco, Google, HP and the others technology companies have home in the same region becomes of relevance. The territorial needs and peculiarities do not draw attention; the proximity helps to perceive more rapidly the buyer needs and the new operating possibilities. For that reason the CT facilitates the aggregation of companies that work in the same sector: principle in total disagreement with the relations among companies planned by SD, that favour the cooperation in different sectors to allow the exchange of wasted material, energy and information and enforce the entire system. If the companies work in the same market it is impossible to design efficient flows where the output of one company becomes input for another. In addition, more the companies are different more the entire system is dynamically stable. If one relation between two different companies has lost its purpose, e.g. because of changed market conditions, there will be more possibilities for these companies to find other kind of relations without failing. The companies in CT are in competition and clusters tend to accentuate it because it improves productivity through improved access to specialized suppliers, skills and information, innovation has great importance as the need for the improvement in productive processes and clusters grow for the creation of new firms and the entrance of new suppliers. However, competition and cooperation in clusters can coexist because they are on different dimensions or else because cooperation at some levels is part of winning the competition at other levels (Porter, 2000). In SD theory the result of relationship among companies is only cooperation, evolving in a new market model, where any business predominates on the other, but each one exists thanks to the others (Bistagnino, 2009).

Less marked differences can be noticed between IE and SD (*figure 3.2.II*), because the fundamental paradigms are closer: industrial system should mimic the nature, planning new model

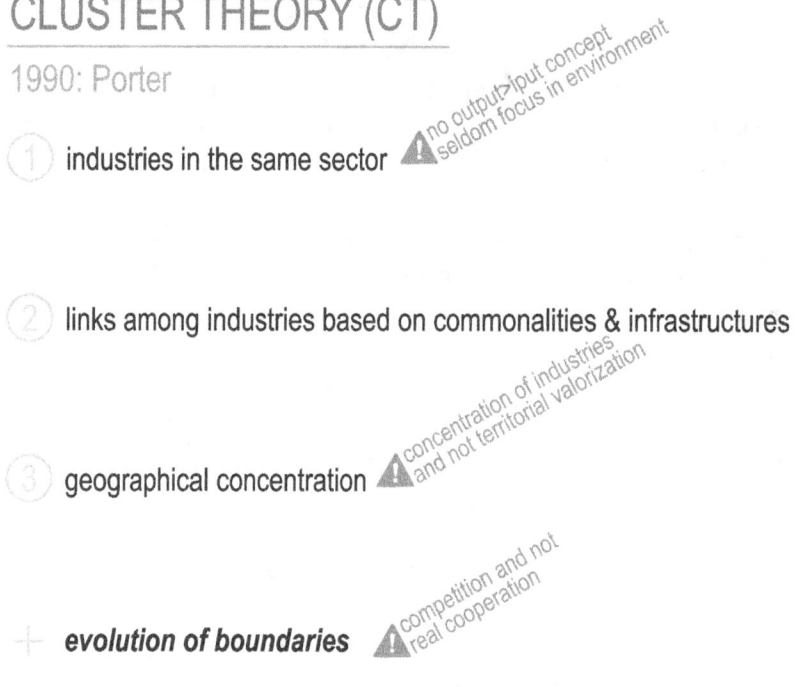

figure 3.1.I: differences between CT and SD.

INDUSTRIAL ECOLOGY

1989: Frosh & Gallopoulos

(1) balancing input/output to the contraints of natural system

(2) dematerialization of industrial output

(3) improving the efficiency of industrial processes

(4) symbiotic links among industries ⚠ competitiveness ⚠ symbiotic link and not cooperative interaction
    cooperation

(5) creating new structures

(6) industrial eco system

+ **close loop** ⚠ close loop and not open systems
  cycle

+ **holistic view**

+ **flows of** — **material** ⚠ chemical integrity ⚠ flow of information
               energy

+ **benefits** — **social**
                economical ⚠ broader framework of community sustainability
                environmental

figure 3.2.II: differences between IE and SD.

for human being. In industrial systems, the waste produced by one company would be used as resources by another; therefore, no waste would leave the industrial system or negatively impact the natural system. The use of products apparently worthless or of endowing bodies with properties renders them of increased value to industry. This fundamental and basic tenet is synthesized in SD with the *output>input principle*; more details are added in IE with a dematerialization of output and an improving of efficiency in industrial processes.

These theories have other similar principles:

- the holistic view of industrial production and environmental protection, expanding the boundaries beyond the scope of single process and core business;

- the results of economical, environmental and social benefits;

INDUSTRIAL SYMBIOSIS

2000: Chertow

(1) output > input

(2) **symbiotic links among industry**
mutualism and commensalism ⚠ *cooperation*

(3) self-organization

(4) geographic proximity

flows of
- material
- energy
- water
- by-products ⚠ *information*

benefits
- economical
- environmental
- social

figure 3.2.III: differences between IS and SD.

- the creation of new structures, enabling continuity and networking, in IE changed into a real autopoiesis of the systems, and reproducing automatically themselves in SD;

- the generation of industrial eco-system where the companies can work closely to one another, stated by IE, finds more emphasis in SD with a territorial planning, so much that a project cannot be exported tout court in another context.

- the consideration of information flows at physical and chemical levels to design new industrial systems; those allow the connections among parties and control the entire system[13].

However, these theories present some peculiar differences:

- in IE, the relations among industries are symbiotic, where a single-industry dominates clusters; in SD, they are cooperative and equal, without supremacy.

- in IE, the loops are closed to recycle the waste in the system, finding uses for waste flows from industrial process or re-engineering material processes to generate usable waste and recyclable products; in SD, the systems are open to metabolise the material in other industrial processes that are different from the initial ones, considering that also in nature one natural reign cannot digest its own waste.

- in IE the intensive use of waste for by-product manufacture increases the force of competition, that can negatively affect the individual, but can positively affect the system. In SD the companies work in a situation of cooperation because the ultimate goal is not the economical preponderance, but the benefits for both people and environment. The first SD principle (man at the centre of the project) states the philosophical base of this theory.

The considerations convey for IE in comparison to SD can be done also for IS, being part of IE (*figure 3.2.III*). In IS, the relations among companies are always symbiotic with the mutualism and the commensalism, that gives the sense of help from one stronger company to another one dependent on it. Furthermore, IS does not consider the exchange of information and know-how, that creates a material culture connected with the territory.

In conclusion, the SD gives a methodology that synthesizes the previous theories with 5 principles directly applicable in real situation to design complex systems aiming at zero emissions. The multi-disciplinary vision brings together all different kinds of scientific knowledge for an innovation inspired to the real dynamics of nature. This renovated flowing generates a new economic model which, enhancing the local resources, gives new life and revival to the territorial and cultural peculiarities.

---

13 The ecosystems can be considered cybernetic, but the control capacities are internal and diffuse in the system.

3.2.b Consistency

Researches who use a systemic approach have a significant interest in leverage points or places within complex systems where a small shift in one parameter can have large impacts (Meadows, 1999). The ways to intervene in a system, for increasing its effectiveness, are the gain around driving positive feedbacks, the structure of information flows (who does and does not have access to what kinds of information) and the power to change, evolve or self-organize (system resilience). Leverage points are accessible with rigorous analysis, but the advantages of flow studies are considerable:

- large complex systems design, where many different systems levels are considered;

- economic consistent flows;

- evaluation of the present, suggesting last-long strategies;

- pro-activity in future problems.

However, these advantages have also some obstacles to be taken into consideration:

- generic environmental pressure often does not have specified risks;

- disregarding in manage physical throughputs;

- uncertainties;

- demand of resources relied on expertise knowledge.

SD needs to move into a phase where its methodology could be tested in different industrial sectors with real life cases and to make use of the outcomes. This research is intended to contribute to SD methodology, mainly through attempts to test the practical applicability of the principles into operational energy production and through critical thinking and re-conceptualizations of the methodology in the energy sector. Such contributions are intended to assist the refinement of relevant practical and academic work.

The analysis of complex systems by scientific means with a societal relevance supports decision making, this approach turns environmental problems into business opportunities, so companies can transform unwanted waste and by-products into valuable products. The use of natural resources and waste is a significant cost for many companies, which could be reduced through precautionary measures and improved sorting.

3.3 When

The acquired knowledge on SD comes from former studies[14] and it hass been growing up along these last three years. Particularly, the first year I have acquired a strong awareness in the theory and the ability to practice the tools. Thanks to this solid background, the theoretical aspects of this research looked at the connections of SD theory with previous methodologies in dynamic industrial management. Environmental studies are the basic concepts to understand interactions between society and production, which includes multiple disciplines, such as technology, material, history and physics. This encourages a more holistic perspective with the use of knowledge from different disciplines and with the maps of interactions between elements in bioenergy systems. This multi-disciplinary is valuable in research methodology because it gives a different perspective on the issue that usually is in the hand of engineers and economists. Furthermore, during that time of studying and reading, attending the III level classes has been foundamental, talking with experts in and outside academia, and taking part in international conferences.

The second year was mainly based on field research that in any case involves also some theory. Once the interaction with the real life cases has started, the research focus was dominantly focused on what is believed to be the bottleneck by the time. In light of the improved understanding, the results become evident and the framework designed.

However, in the third year the SD theory took again a preponderant rule. The cross analysis among the different case studies and between the theoretical/practical parts allow to design the framework for LED. Research activities are shaped by research paradigms from the practical development to how findings are analysed and presented, in a constant reflective and iterative process on alternative interpretation of the data and findings (Hammerskey, 2000).

3.4 Where

In **Italy** the **SD** finds its actual theorization and practical experimentation in the design at **Politecnico di Torino** (Bistagnino, Lanzavecchia, 1998): it attends to ecocompatibility of industrial products and processes. In late 90s he starts to cooperate with the Zero Emissions Research and Initiatives Foundation in Switzerland (Pauli, 1994), to merge the economical and technological knowledge with the design and environmental one. The result was a method with clear and specific principles that can be applied in concrete fields (from agroindustry to events) as shown in Bistagnino's last book "*Design Sistemico. progettare la sostenibilità produttiva e ambientale*" (Slow Food Editore, 2009). However, in 50s and 60s the **Hochschule für Gestaltung** in **Ulm** (**Germany**) was the first school of design that had a interdisciplinary approach. Under the leadership of Maldonado, the school dropped the artist focus of Max Bill and proposed a new philosophy of education as an "operational science", which embodied both art and science. This approach was strongly based on aesthetics: the *gestaltpsycologie* is based on the perception as integrated structure (not as synthesis of isolated elements). The singular elements acquire different senses depending on the made up units. (Krampen & Hörmann, 2003)

A part from design field, the main features of systemic thinking have been taking shape in the early XX century, improved thanks to different disciplines like quantum physics, ecology, cybernetics, and economy. At **Yale** in **New Haven**, the embryologist **Ross Harrison** states the importance of replace the concept of function with that one of organization. The two main aspects of organization are the configuration and the relationship: the schema is a configuration of ordered relations (Capra, 1996). The essence of this new reality moved from the elements to the relationships. Furthermore, in the Yale University's School of Forestry & Environmental Studies etablished the **Center for Industrial Ecology** in 1998, thanks to **Marian Chertow**.

At the **University of Jena** in **Germany**, the ecology, term minted by **Ernst Haeckel**, brought two important concepts in the systemic thinking: community and web. The ecological communities were considered as one organism composed by parts that depend on each other. The concept of web was born in the 20s because of studies on food chains and metabolism that became food web.

Before the end of the 30s, the main principles of systemic thinking were stated but the scientific movement was recognized at the end of the 40s by the work of biologist **Ludwig von Bertalanffy** at the **Stanford University** in **Palo Alto**. He stated that the living organisms are not closed systems, as thermodynamics said, but open at energy level and close at organizational level.

The **cybernetics** studied how living systems and machineries transform fundamental information to do their activities. This theory, developed by mathematicians **Claude Shannon** and **Warren Weaver**, was based on the fundamental principle that the information passes through signals that have many shapes as configurations. Afterwards, the cybernetics was interested about the consequences of the information on the machinery behaviours (Wiener, 1948).

In the 50s and 60s, the systemic thinking has a great influence on engineering and economical sciences, so new disciplines were born (systems engineering, systems analysis and systemic management).

By the time practical experiences in the field of industrial processes were put into practice, it already was the 70s. Some managers began to use the systemic approach to solve problems in the industrial organizations. The results were the first theorization in models, like the Systems Dynamics by Jay **Wright Forrester** in the Sloan School of Management at the **Massachusset Institute of Technology** in **Boston** and the Management Cybernetics by Stafford Beer.

In the 80s, at the **Santa Fe Institute** in **Santa Fe**, **George Cowan** studies nonlinear dynamics, including what is known as Chaos Theory. Mathematical equations were used to predict the behaviour of systems, such as insect colonies, human societies and cultures, and economic interactions. For the past decade, Cowan has used neuroscience to study the relationship between physiological changes in children's brains and their behavioural development.

The new business model proposed by SD booms in the philosophical concepts of **Fritjof Capra**, as he teaches in the **Centre for Ecoliteracy in Berkeley** (USA). The institute supports and advances education for sustainable living, because of the pivotal role of pupils in moving beyond the growing environmental crisis and toward the sustainable society.

In **Pauli**'s last book "*The Blue Economy: 100 Innovations inspired by nature that can generate over decade 100 million jobs*" (Paradigm Publications, 2010), there is a series of new businesses inspired by the Nature. The **Blue Economy** permits to respond to the basic needs of all simply with what exists. As such, it stands for a new way of designing business: using the resource available in cascading systems, where the waste of one product becomes the input to create a new cash flow.

The cooperation of these institutions all over the world helps to spread the systemic thinking in many different fields. For example in Europe, the **International Institute for Industrial Environmental Economics** (IIIEE) **at** Lund University in **Sweden** and the **National Industrial Symbiosis Programme** (NISP) and **University of Hull** in **England** work on IS in industrial management field. In the IIIEE, the improvement of resource use efficiencies through inter-organisational actions, in geographical localities, and its major objectives are: identifying the characteristics of IS networks and catalysing the development of such networks in different European regions. The NISP operates at the forefront of IS thinking and practice, helping companies take a fresh look at their waste>resources with twelve regional teams.

3.5 How

How the SD is used in that research: in the **theory for planning** and in the **practice for reflecting**. To serve the dual purpose of being part of doing research and intervention at the same time: this research faces the challenge of being able to get sufficiently involved in action and stand back, when necessary, from action and reflect on it, to add up to the theory and body of knowledge. Utilizing a combination of *desk* and *field research* allows a more in-depth understanding of reality from different viewpoints, which is crucial when exploring topics or issues involving a range of actors, like the energy sector. This research is heavily focused on qualitative methods, including literature reviews, case studies, site visits, stakeholder interviews, industry interactions and international agencies connections. The foundation of this research is therefore the use of diverse kinds of data sources and a mix of qualitative methods. The *desk research* detects the information already written by others, in an important identification of the sources with their reliability. The *field research* directly observes the reality by empirical and personal experience. That means the *desk research* is less expensive and also less original; but it helps to define the headway processes of the research. **The combination of *desk* and *field research* guarantees a rich and deep understanding of the facts to define an original framework.** Depth and richness of sources are fundamental for a good result in research and design. The case history gives data that need to be delved by parameters comparable with other standardized cases. This kind of analysis needs the cooperative work of experts with different specializations. The result is a strategy with multi-actions to supervise the feedback and reach the goal. In that process, typical in living complex systems, there are many opportunities to calibrate the standard model with the competences of experts, related to their knowledge and direct experience. The experience of the researcher is found in the quantity and quality of *"super-constants"* through case studies (Celaschi & Deserti, 2007). This research is built upon showing the European energy situation, synthesising the information from relevant literature about SD and eight case studies. The need for a better understanding regarding SD concepts contributes to intensify local sustainability and to test it. Such understanding is particularly important to give a new direction to SD work that can potentially catalyse more profound changes. In this study, such understanding was sought through studying selected production systems that resulted in useful lessons. These systems were selected because those systemic relations are operational, or planned, in particular the *output>input principle*. The SD methodology is important to planning because it underscores a **key-role for designers**: the crucial link between productive and social sustainability. Indeed, the fields of planning and sustainability share similar characteristics: they embrace a **system-oriented, holistic, integrated approach** to consider diverse needs in promoting public welfare.

Designers are in a unique position to assess how SD could fit within an overall strategy for sustainability in their community. However, the designer's role implementing systemic planning is still emerging. Fritjof Capra states designers need to be ecoliteralized.

First, **planning can provide a comprehensive perspective** of ways that SD could contribute to community sustainability. Designers regularly study the industrial metabolism, integrating issues such as natural resources, infrastructure, land use, economic development and equity. Designer can identify potential linkages, analyse and provide information on potential benefit and costs, and develop an organizing framework for implementation.

Second, **designer can co-ordinate and communicate with multiple sectors of society** for implementing SD. Successful implementation depends on productive partnerships between industry, government and the community (Cosgriff Dunn & Stainemann, 1998).

Third, designers' work with regulatory administration can help to **position systemic projects within the community**. Designers are often directly involved in the development and implementation of regulation, as well as the identification and creation of incentives.

Fourth, designers' regular involvement with the public care can help to **carry plans into actions**. Increasingly, planning involves politics. Designers could link partners and community, which could help to establish co-operative relationships and resolve conflicts. Public support is crucial, and an educated, involved community can offer useful suggestions for improving the implementation process.

For designing complex systems, the ability to iterate rapidly is especially important since design teams need to assess the robustness of alternative designs under a variety of different scenarios and assumption. An **open dialogue among the interested parties** is necessary to establish agreement on technical, legal and ethical conditions; history has shown that it is wise for designers to consider not only the product technology but also the socio-economic system within which it will be introduced.

CASE STUDIES ANALYSIS: PRACTICE

4.1 What

Given the nature of the research questions, the method chosen in this thesis work is a case studies analysis, which addresses the technical parts of the systems as well as the environmental, economical and social ones. It is equally important to understand and address the technical aspects of the systems as the other aspects, with a fair balance among the parties. The multi-disciplinary character of that research, utilising concepts and knowledge from different disciplines, makes it difficult to satisfy experts in specific disciplines (McCormick, 2007). The analysis of these dimensions can be qualitative and quantitative, so the case studies have been examined as cross-level and cross-scale phenomena.

The case studies methodology is appropriate when *"what", "why", "when", "where", "how"* questions are asked, but not statistical results can be obtained nor can the results be directly generalized to other cases outside of the study, for that reason the cross analysis is needed.

A total of **eight case studies** have been explored and grouped in two main categories (*figure 4.1.I*): the macro systems and the micro systems. In the first case, they are large-scale plants that provide energy to entire cities (i.e. *Växjö*) or district (i.e. *Vignolo*); in the second case, they are small-scale plants that supply just a factory (i.e. *Bioagro*) or a single house (i.e. *Villa Ödman*).

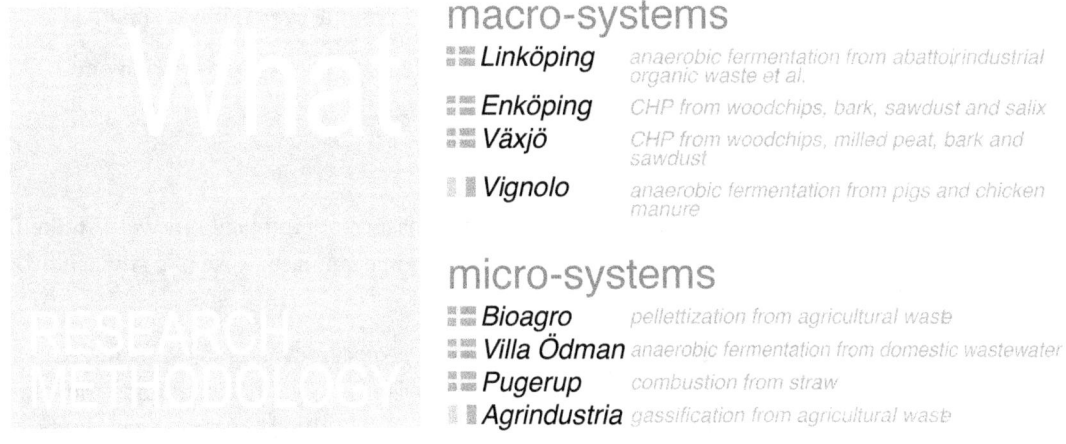

macro-systems
- **Linköping** — anaerobic fermentation from abattoir/industrial organic waste et al.
- **Enköping** — CHP from woodchips, bark, sawdust and salix
- **Växjö** — CHP from woodchips, milled peat, bark and sawdust
- **Vignolo** — anaerobic fermentation from pigs and chicken manure

micro-systems
- **Bioagro** — pellettization from agricultural waste
- **Villa Ödman** — anaerobic fermentation from domestic wastewater
- **Pugerup** — combustion from straw
- **Agrindustria** — gassification from agricultural waste

figure 4.1.I: macro- and micro-systems selected in the research.

4.1.a Macro-systems

The case studies analysis consists of mapping energy and material flows within their territories and influences. For the macro-systems, the areas of influence cover entire cities (*figure 4.1.II*).

The city of **Linköping** has a complex transport system totally fuelled by biogas. The anaerobic fermentation plant processes the abattoir waste of the municipality and other organic industrial waste to produce biogas (7,700 m^3/y) and fertilizer (52,000 tonnes/y). The actors involved in the projects are the municipal power organization and transport company, the municipality, the university and the food companies, like the Swedish Meat AB and the farmers cooperative.

The city of **Enköping** has a 100% biomass (woodchips, bark, sawdust and salix) fuelled energy system with synergistic connection between the power plant, the local farmers and the municipal wastewater treatment system. The mixture of bottom ash from the bioenergy plant and digested sludge from the wastewater treatment facility is used as fertilizer, that is delivered free of charge to salix farmers. The remaining ash is used as terracing material to cover waste deposit at the Annelund Dump. The main products of this steam boiler are heat (230,000 MWh/y), used to warm up 1,200 houses, and the electricity (95 MWh/y).

Växjö Municipality has installed a power plant totally fuelled by biomass (woodchips, peat, bark and sawdust). The main products of this steam boiler are heat (66 MWh/y) for the district heating and the electricity (36 MWe). Furthermore, it produces also fly and bottom ashes, the former is used again inside the process and the latter is used for fertilizing the soil.

Vignolo is the site were the private company Marcopolo Engineering decided to settle a biogas plant in Piedmont (Italy). The anaerobic fermentation processes 100 tonnes per days of cow and chicken manure to obtain biogas (2.4 M m^3/y), electricity (7,000 MWh/y) and bio-fertilizer. The power plants sell the produced green energy to the national electric grid, but not to the neighbouring homes. it cannot provide directly the houses around, even if it is estimate.

These macro systems are explained in detail in *Annex A*, dedicating one chapter each one.

4.1.b Micro-systems

In the case of small power plants, the processes and their associated implications were studied, including those on the surrounding resource flow dynamics, even in the case of a single house (*figure 4.1.III*).

In the city of **Östra Tommarp**, there is the Bioagro project developed by Skånefrö AB and Ecoera AB, with the cooperation of the Chalmers University of Technology, the biomass suppliers and the machinery manufacturers. The agricultural waste, like sawdust, bark, straw, hemp and switch grass, are pelletized and then burned to obtain heat (1.5 MWh), electricity and ash. Now they are improving the use of ash with a pyrolysis pro-

figure 4.1.II: localization of the four analyzed macro-systems.

cess to have a good fertilizer like the biochar to produce biochar (an organic fertilizer).

In Kvissleby hamlet, the architect Anders Nyquist designed a single house with a very innovative and complex system to produce heat (40 KWh), that is enough for the internal uses because of efficiency, fertilizer for the garden and clean water. **Villa Ödman** uses the Splitbox technology in a very holistic way, obtaining environmental benefits and great comfort.

In the city of Höör, the **Pugerup** farm installed a straw boiler to processes all the agricultural waste that it produces, so mainly straw and small quantity of woodchips. The farm activities include grain production, forest cultivation and pig breeding. The boiler produces steam for heating (600 KWh) the entire farm, included the pigsties.

In Piedmont (Italy) a small enterprise, **Agrindustria**, has transfromed agricultural waste in industrial by products since the '80s. Now it has decided to use some of their biomass to produce green energy. The gasification plant is fuelled by woodchips, under wood, sawdust and waste wood from boxes and pallets. It produces heat (800 KWh), that is directly used in the internal industrial process. Soon it will warm up the offices, and sell electricity to the national grid with green certificates. The biochar, that produces will be used to fertilize the energy crops.

4.2 Why

The motivations in choosing these **eight case studies** stay in the **approach**, the **size**, and the **experiences** (*figure 4.2.1*). The selection of case studies is based on:

- the production bioenergy[1];
- the complexity they have with more or less large networks of organizations;
- the compliance with the SD theory, above all the output>input principle.

The different size of plants allows understand the regional effects with macro systems and the local effects with micro systems. In that way, the limits in using local resources to produce green energy become evident.

The localization of the cases is divided in two specific countries of EU: Sweden and Italy. The first country was chosen for its historical experience in bioenergy and holistic approach to the field. Italy was chosen because of its relationship systems design theory and to help the country reach the European targets in the production of energy from renewable resources.

4.2.a Approach

When one observes the statistics and cases of everyday reality it is not hard to notice that even companies with an excellent background in science and technology, which already conducted innovations and projects with great success, were not always able to get profits out of their innovative activities. The evolution of technology shows that technological excellence by itself is not sufficient to impose the spread of a product, that imitators can be more successful

---

[1] When biomass (plants and animal matter) is utilized to produce heat, electricity or fuel for tran sport it is commonly called bioenergy.

figure 4.1.III: localization of the four analyzed micro-systems.

than their innovators and hence extract profits from the innovator's gains, that brilliant ideas are often neglected for years before being economically exploited, and that many other factors could help to determine the managerial and economical failure of the company's technological innovations (Calderini et al. 2003). So the selected cases are chosen not for their technological contribution, but for their holistic approach and their effort to consider complex solutions.

In these cases, it is possible to reflect the SD principles even if they are not designed with a conscious use of that methodology. In that way the methodology enforces itself because it can verify the successful cases. The best practices can acquire a method to easily increase their benefits. The dual benefits have further promotion with the test in a micro system: Agrindustria. That case study became a real test bed of the designed framework, so much that the theoretical and the practical parts grew up together with multiple and continuous exchanges.

The analysis of all case studies considered:
- the costs involved in energy system planning;
- the structure of the energy markets;
- the economic assessment of energy systems:
- the multi-generation perspectives in energy-related markets.

4.2.b Size

The large-scale units generate centrally and transmit bulked power (heat, electricity or biofuel for transport). That is the most common model with centralised generation and distribution through a transmission network (the grid).

The small-scale power plant are often described as "distributed", "decentralised", "modular", "localised" or "embedded", because they produce energy in relatively small amounts near the consumer with higher degree of flexibility (Madlener and Wohlgemuth, 1999).

The different size of plants allows to understand the regional effects with macro systems and the local effects with micro systems. The dimension of the plant is obtained by different tech-

approach
- *production of bioenergy*
- *complexity of networks*
- *compliance with SD*

size
- *macro-systems:* regional effects
- *micro-systems:* localt effects

size
Sweden: - historical experience in bioenergy;
- holistic approach to the field.
Italy: - relations with SD;
- help it to reach the EU target.

figure 4.2.I: motivations in selection of the eight case studies.

nologies and influences different area (macro systems collect the needed biomass in a round about 100 km, micro systems in at least 40 km). In that way, the limits in using local resources to produce green energy become evident. A truly sustainable development may be achieved with the diversification and localization of energy sources and systems if the adverse impact of each energy system is sufficiently small and well fit within the tolerance limit of the environment. The combination of energy resources to attain the optimal energy supply may vary for different territorial regions to suit the local situation, which is influenced by local practices, current resource level and variety and environment. Clearly, the territorial approach encompasses the benefits of economy-wide, sectoral and local approaches, and more. The limit conditions could change, but the system strictly connected with its territory defines, in autonomous ways, new balances inside the new limits. Working with a clear limit is very effective in achieving the desired goal.

Furthermore, the consideration of two different dimensions provides a conceptual and empirical formulation of the mechanism that could link these macro-micro systems. Comparison processes advance as one route through which recurring factors can be connected. The relation between small- and large-scale power generation can give a significant contribution to define new modular generation technologies.

4.2.c Experiences

The geographical aspect of that project highlights two main areas: the Nordic peninsula and Italy. That situation is useful to reach the goal of that project because the practical experiences in the field of green energy, typical in Northern Europe, should be merged with a more theoretical and methodological aspects, stronger in Italian university. For example Scandinavia as a whole is a leading region in this area, both in terms of the contribution bio-energy makes as in terms of technological developments. Germany, the Netherlands and the UK are countries with considerable ambitions in the bio-energy field, but with the natural 'handicap' that the national biomass resource base is less abundant than in Northern Europe. France and Spain are key examples in Southern Europe (Faalj, 2004); the aim of that research is to add Italy in this scenario.

In Sweden, saving and efficiency improvements have been important parts of the Energy policy since the first oil crisis in 1973. Fossil fuels carry a CO_2 tax and energy tax in Sweden, which has forced a restructuring of the energy sector from using fossil fuels to renewable biofuels from the forest and arable land. Municipalities have historically had a strong political position, which express through local self-government. Among else, Swedish municipalities are often responsible for energy planning, electricity, sewage treatment and fresh water supply, public transportation and environmental protection. Often, the energy activities of the municipalities are handled through municipally owned companies. In Italy the Government made an effort with the 2007 financial bill. The recent Italian Law 99/2009 provided for the publication of an

Extraordinary Plan for Energy Saving and Efficiency, to make up the gap with other EU Countries. Actually the support mechanisms are ensuring remuneration for investment in various renewables energy and energy efficiency operations and encouraging the growth of related industries. The incentive schemes for electricity produced by plants using renewable sources are based on the green certificate (with a minimum quota of new electricity production). Furthermore, there is a fixed all-inclusive tariffs for electricity fed into the grid by renewable energy plants with a maximum power output of 1 MW and for photovoltaic and solar thermodynamic plants a feed-in tariff mechanism.

The stronger experience of Sweden in improving green energy power plants is merge with a stronger methodology of Italy in SD theory. These two different cultures and background empower each others and generate a tested framework that will help the EU to reach the targets of increasing the share of renewable resources to produce energy.

4.3 When

When the case studies are analysed is mainly during the second year of that research after having acquired a strong awareness in the theory and the ability to practice the tools. Once the interaction with the real life cases has started the research focus was dominantly focused on what is believed to be the bottleneck by the time. In light of the improved understanding, the results become evident and the framework designed.

The first year of that research was mainly dedicated to acquire the theoretical background on SD, that has former knowledge thanks to I level master in Ecodesign and II level international master in SD. Thanks to this solid background, the theoretical aspects of this research solve the connection of SD theory with previous methodologies in systemic industrial management. That background forms the foundation of this thesis work. Environmental studies are the multidisciplinary examination of interaction between society and production, which includes disciplines, such as technology, material, history and physics. It is important to acknowledge the skills and the knowledge because the research methodology reflects on that basis and give a different perspective on the bioenergy field, usually is in the hand of engineers and economists. Furthermore, during that time of study research by studying and reading, was fundamental attending III level classes, talking with experts in and outside academia, taking part in international conferences. These activities have shaped this research and given the opportunities to meet experts in the bioenergy field and to access data and documents.

The second year was stated by field studies, going to the International Institute for Industrial Environmental Economics at Lund University in Sweden that give a practical support to that thesis. The main orientation of the energy work at the Institute is the evaluation of alternatives related to, and policies for, Energy for Sustainable Development. The selection of case studies was assisted by the critical analysis of European context in the field of green energy production, in particular in Sweden and in Italy. All the selected cases were visited in person to interview various informants from the communities, the industries, the governments and the academia. Multiple viewpoints check "facts" and identify issues of disagreement or tension (Morrow and Brown, 1994). The same questions and themes were discussed with different informants. Such an approach was vital to the reliability and validity of the data collection process (Bloor, 1997). The third year was mainly dedicated to the cross analysis among the different case studies and between the theoretical and practical part to **design a framework for LED by means of SD**. Research activities are shaped by research paradigms from the practical development to how findings are analysed and presented, in a constant reflective and iterative process on alternative interpretation of the data and findings (Hammerskey, 2000). Furthermore, it was implemented with a test bed in one of the micro system analysed, effecting a real action research (Ottosson, 2003). The combination of action and change with research produces a flexible research made of planning, acting, observing and reflecting, with a complete involvement in the project of Agrindustria power plant. Recent focusing on action research arises partly from the desire to go beyond identifying and describing problems, to engaging with actors and to find solutions. In other words, to have a real impact on those actors involved in a study. This greatly improved the understanding of "real-life" systems and helped to focus the research on applicable solutions. The test bed in Agrindustria micro-system "leap-frogs" certain problems to framework formulation and implementation.

4.4 Where

Where the case studies are settled was clarified above with the **experience motivations** (in Sweden and in Italy) to have a complete scenario at theoretical and experienced levels. The research joins the **experiences in the green energy production from Nordic Countries** with the **methodology in SD from Italy**, to promote new and renewable energy sources, from biomass residues, for decentralization production of electricity and heat and its diversification. The **localization** is one of the most important aspects of that kind of projects, because the bio-energy systems are influenced by their context in terms of natural resources and socio-political issue (as one of the SD principle said: act locally). Thus, at the beginning of each case study the context is analysed with some **maps and sketches** to understand where they are settled, in a sense of **geographical condition**, **land use** and **connections**: a reworking through **conceptual abstraction**[2]. The sources of these information are *Google Hearth* for data, and the *European Environmental Agency* with *Corine Land Cover 2000* for the land use.

In addition, the research takes in consideration the **general different conditions** where the two countries are working on and which are their actions plans for the future. Thus, **politi-**

---

2   That part was developed in cooperation with Fabiana Zanoni, expert in policy making and environmental/territorial planning (knowledge, interpretation of territorial public action) by University IUAV in Venice (Italy).

cal, economical and **social conditions** are described below to understand the differences in context and approach (*figure 4.4.I*). The information are mainly elaborated from the *National Action Plans for the promotion of the use of renewable energy in accordance with the Directive 2009/28/EC* and the *Commission Decision of 30 June 2009* give to the Energy European Commission by all the EU Member States.

Sweden has for many years had a policy for the promotion of renewable energy as part of an overarching policy for sustainable development and efficient use of resources. Even though policies for the promotion of renewable energy are not in themselves an innovation, new components with binding European targets exist. Sweden has had targets in the past and currently has targets, more stringent[3] than in the past.

The Swedish Parliament has decided that the **proportion of renewable energy in 2020 will be at least 50%** of the total energy usage. In order to achieve that **target,** the Government has put forward a number of proposals, including the further development of the electricity certificate scheme for renewable electricity generation. General **economic instruments**, such as carbon dioxide tax, international emissions trading and certificates for renewable electricity are fundamental to its long-term energy policy. The economic instruments are gradually developed and exceptions limited as far as possible, taking into account the risk of carbon dioxide leakage and the competitiveness of Swedish trade and industry. The specific measures and plans to promote the energy from renewable sources are summarized in the *table 4.4.a*. The instruments must be complemented by both **technology development initiatives** and **information** to remove institutional obstacles to renewables. From 2009 the Swedish Energy Agency will make available just over 110,000 M€ per year for energy **research**. In addition to this, in the energy sector the annual grants to universities and technical colleges will increase by 5.5 M€ on in 2010, by a further 5.5 M€ in 2011 and by a further 6.5 M€ in 2012.

The financing is being targeted at the following **areas**:
- large-scale renewable electricity generation and its integration into the electricity grid;
- electric propulsion systems and hybrid vehicles;
- waste refineries;
- biofuels;
- renewable materials;
- new nuclear technology;
- carbon dioxide capture and storage.

Italy aims to **redress the balance of its energy mix**, which is currently too dependent on imported fossil fuels. This process will also involve significant measures to relaunch the use of new-generation nuclear power

---

[3] The governement bills 2008/09 state the 162 and 163 on an integrated climate and energy policy (om en sammanhållen klimat- och energipolitik).

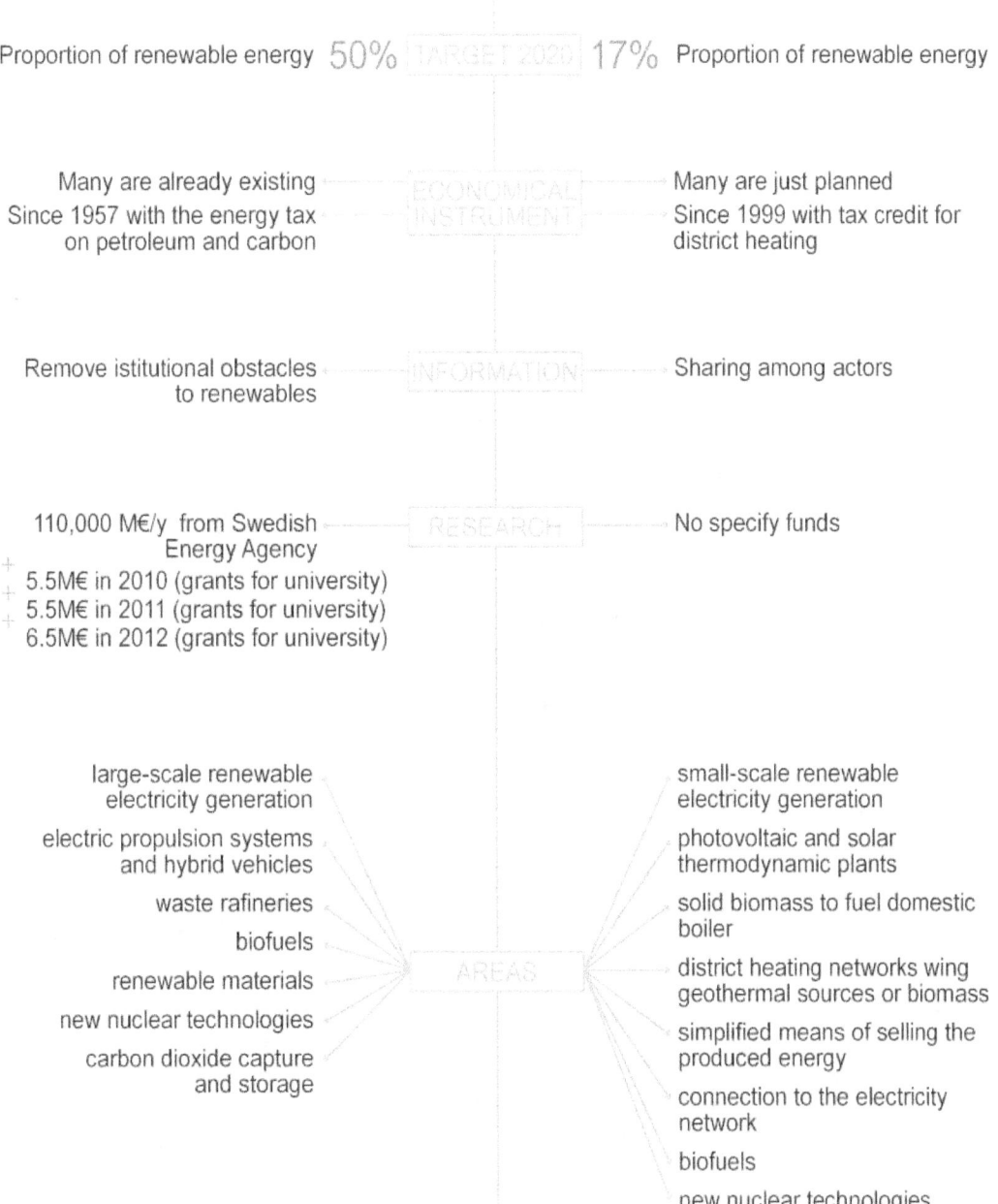

figure 4.4.I: comparison between Swedish and Italian Energy Action Plan.

table 4.4.II: overview of Swedish energy plans and measures.

Name and reference of the measure	Type of measure	Expected result	Targeted group and/or activity	Existing or planned	Start and end dates of the measure
Energy tax Lag (1994:1776) om skatt på energi (Act (1994:1776) on energy tax)	Financial	Fiscal and steering tax for more efficient energy consumption and increased share of renewable energy	Households, enterprises	Existing and planned adjustments	Energy tax on petrol 1924, on the majority of other liquefied petroleum products and carbon fuels 1957, on LPG 1964 and on natural gas 1985 -
Sulphur tax Lag (1994:1776) om skatt på energi (Act (1994:1776) on energy tax)	Financial	Environmental tax to reduce sulphur emissions	Industry and power plants, the transport sector, heating	Existing	1991 -
Carbon dioxide tax Lag (1994:1776) om skatt på energi (Act (1994:1776) on energy tax)	Financial	Environmental tax to reduce sulphur emissions	Households, enterprises	Existing and planned adjustments	1991 -
Nitrous oxide tax Lag (1990:613) om miljöavgift påutsläpp av kväveoxider vid energiproduktion (Act (1990:613) on environment charges on emissions of nitrous oxides from energy generation)	Financially regulatory	Miljöstyrande avgift, steering towards reduced emissions, no direct promotion of renewable fuels.	Industry and power plants, at least 25 GWh per year.	Existing	1992 -
Exemption from energy and carbon dioxide tax for CO2-neutral fuels and for vegetable and animal oils and fats and biogas as a heating fuel Lag (1994:1776) om skatt på energi (Act (1994:1776) on energy tax)	Financial	Promotes the use of bioenergy.	Biofuels	Existing	1991 -
Electricity certificate scheme, Lag (2003:113) om elcertifikat (Act (2003:113) concerning electricity certificates)	Financially regulatory	25 TWh new renewable electricity generation (previously 17 TWh) for 2020 (previously for 2016) compared with 2002	Quota-bound electricity suppliers/consumers and producers of renewable electricity	Existing and adjustment of quota levels	From 2003 the increase in ambition relates to the period 2013-2035
EU-ETS, Lag (2004:1199) om handel med utsläppsrätter (Act (2004:1199) on emissions trading)	Financially regulatory	EU-wide instrument à conversion to the use of renewable energy fuels	Plants within the trading system	Existing with adjustment	New period from 2013

Marketing of wind power, Förordning (2003:564) om bidrag till åtgärder för en effektiv och miljöanpassad energiförsörjning (Ordinance on grants for measures for efficient and environmentally friendly energy)	Financial contribution	Development and demonstration support for wind pilot projects; SEK 350 million 2008-2012; Total granted so far just over 400 million 2003-2009,expected to generate 0.95 TWh (production is also eligible for electricity certificates)	Wind power enterprises	Existing	2003-2007; 2008-2012
Planning support for wind power, Förordning om stöd till planeringsinsatser för vindkraft (SFS 2007:160) (Ordinance on support for planning initiatives for wind power)	Financial contribution	To support the planning process	Municipalities, county administrative boards, municipal and regional cooperative bodies	Existing	2007-2009 (there are funds remaining and it will also be possible to apply in 2010)
Investment aid for photovoltaic cells connected to the grid, Förordning (2009:689) om statligt stöd till solceller (Ordinance on government support for solar photovoltaic cells)	Financial	Target is that the number of operators will increase in Sweden, that the system costs will be reduced and that electricity from solar photovoltaic cells will increase by 2.5 GWh during the period	Companies, public and private organisations and private individuals Concerns solar photovoltaic cell systems connected to the electricity grid (also entitled to electricity certificates)	Existing	1 July 2009 – 31 December 2011
Financial support for investment for solar heating Förordning (2008:1247) om stöd för investeringar i solvärme (Ordinance on support for investment in solar heating)	Financial			Existing	2009-2010
Aid for conversion from directacting electrical heating, Förordning (2005:1255) om stöd för konvertering från direktverkande elvärme i bostadshus (Ordinance on support for conversion from direct-acting electrical heating in residential buildings)	Financial	Conversion from direct-acting electricity to district heating, bioenergy and heat pumps	Owners of residential buildings or associated premises	Existing	Funding must only relate to measures that have been commenced no earlier than 1 January 2006 and completed no later than 31 December 2010.
HUS/ROT relief (renovation, maintenance and conversion and modifications), lag (2009:194) om förfarande vid skattereduktion för hushållsarbete, HUSFL (Act concerning the procedure for tax reductions for household work)	Financial	Tax credit for work costs including for investment in renewable energy	Individuals (single family houses and housing co-operatives)	Existing	8 December 2008-30 June 2009, (ROT), 1 July 2009- (HUS)

SOURCE: National Renewable action Plan in line with Directive 2009/281EC, Swedish Parliament (2010)

Obligation to supply renewable fuels (Pumplagen) (the Pump Act), Lag (2005:1248) om skyldighet att tillhandahålla förnybara drivmedel (Act concerning the obligation to provide renewable fuels)	Regulating	All retail outlets (above a certain volume) must supply renewable fuel	Retail outlets for fuel	Existing	2006
Grants to fuel retail outlets for investment in pumps other than ethanol, Förordning (2006:1591) om statligt stöd till åtgärder för främjande av distribution av förnybara drivmedel (Ordinance on government support for measures to promote the distribution of renewable fuels)	Financial	114 retail outlets had been given grants (average just over SEK 1 million per application) in 2009 for the installation of biogas pumps	Retail outlets for fuel	Existing	2007-2009, it is still possible to apply for grants for works that had been commenced before the end of 2009.
Vehicle tax, Vägtrafikskattelag (2006:227) (the Road Traffic Act), and Lag (2006:228) med särskilda bestämmelser om fordonsskatt (Act with special provisions concerning vehicle tax)	Financial	Environmentally-steering	Vehicle owners	Existing	Enhanced environment control introduced 2006
Exemption from vehicle tax for environmental cars, Lag (2006:228) med särskilda bestämmelser om fordonsskatt (Act with special provisions concerning vehicle tax)	Financial	Promoting environmentally-friendly cars	Vehicles owners, the vehicle industry	Existing	2010, retroactive from 1 July 2009 - 2012
Reduction in the amount of benefit for environmental cars, inkomstskattelagen (1999:1229) (the Income Tax Act) and Skatteverket's (the Swedish Tax Agency) regulations and general guidelines	Financial	Promoting environmentally-friendly cars (compare the taxable benefit of environmentally-friendly cars with equivalents alternatives, even though the environmentally-friendly car is more expensive to purchase)	The company car sector	Existing	2009 -
Environmental cars in government procurements, Förordning (2004:1364) om yndigheters inköp och leasing av miljöbilar (Ordinance concerning procurements made by authorities and leasing of environmental vehicles), Förordning (2009:1) om miljö- och rafiksäkerhetskrav för myndigheters bilar och bilresor (Ordinance concerning environmental and traffic safety requirements for authority vehicles and journeys)	Regulating	Promoting environmentally-friendly cars	Governmental authorities	Existing	1 January 2005 -

Government public procurement with environmental requirements, Lag (2007:1091) om offentlig upphandling (Public Procurement Act), lag (2007:1092) om upphandling inom områdena vatten, energi, transporter och posttjänster (Act on public procurement of water, energy, transport and postal services)		Promoting the development of new climate-efficient technologies	Governmental authorities	Existing	
Investment support for biogas and other renewable gases, Förordning (2009:938) om statligt stöd till åtgärder för produktion, distribution och användning av biogas och andra förnybara gaser (Ordinance concerning government support for measures for the production, distribution and use of biogas and other renewable gases)	Financial	Funding for projects that contribute to increased generation, distribution and use of renewable gases	Production centres, distributors and consumers of biogas and other renewable gases	Existing	1 November 2009 - 2011
Investeringsstöd för produktion eller förädling av biogas inom Landsbygdsprogrammet (the Swedish Rural Development Programme), Förordning (2007:481) om stöd för landsbygds-utvecklingsåtgärder (Ordinance concerning support for rural development measures)	Financial		Farmers and other rural entrepreneurs	Existing	
Investment support for planting energy forests on arable land within Landsbygdsprogrammet (the Swedish Rural Development Programme), Förordning (2007:481) om stöd för landsbygds-utvecklingsåtgärder (Ordinance concerning support for rural development measures)	Financial	Target regarding multiannual energy crops that an area equivalent to 30,000 hectares is to have been planted during the programme period (2007-2013)		Existing	
Support for climate and renewable energy projects, special funds allocated within Landsbygdsprogrammet (the Swedish Rural Development Programme), Förordning (2007:481) om stöd för landsbygds-utvecklings-åtgärder (Ordinance concerning support for rural development measures)	Financial		Company and project funding	Existing	2010-2013
Support for energy identification for SMEs, Förordning (2009:1577) om statligt stöd till energikartläggning (Ordinance concerning government support for energy identification)	Financial	Funding for energy surveying in companies that have an energy consumption in excess of 0.5 GWh, up to a maximum of SEK 30,000 per enterprise	Small and medium-sized enterprises (energyintensive enterprises are included primarily of PFE) and certain agricultural enterprises	Existing	2010 -

The **target** for Italy is that the 17% of final energy consumption must be covered by renewable sources in 2020.

There are numerous **support mechanisms** (tax relieves) already in operation to make up for the insufficient level of remuneration for investment in the renewable energy and energy efficiency sectors, which has so far been provided solely by **market mechanisms**. The specific measures and plans to promote the energy from renewable sources are summarized in the *table 4.4.III*.

The biomass sector is important, partly due to its specificities, and it will be promoted organically, by identifying measures aimed at increasing the availability and mobilisation of biomass, directing its application not only towards electricity generation but also more convenient forms for covering end-use: heat production to meet heating requirements and for co-generation.

In general, the Italian Government intends to direct and **expand the research and development support schemes**, with the aim of strengthening Italian industry's competitiveness in those technologies and applications for which, given the significant potential for propagation, there is a production structure equipped with the necessary know-how for further development.

As well as promoting renewable sources for heating and cooling and transport uses, the measures to be implemented will principally relate to electricity network management, further streamlining of authorisation procedures and the development of **international projects**. The involvement of and coordination between the various local authorities and bodies will be essential, as will the **sharing of information**.

The principal **plans** to increase the electricity production from renewable sources are:
- small-scale renewable electricity generation and its integration into the grid;
- photovoltaic and solar thermodynamic plants;
- solid biomass to fuel domestic boilers;
- district heating networks using geothermal sources or biomass;
- simplified means of selling the produced energy;
- connection to the electricity network;
- biofuels;
- new nuclear technology.

table 4.4.III: overview of Italian energy plans and measures.

Name and reference of the measure	Type of measure	Expected result	Targeted group and/or activity	Existing or planned	Start and end dates of the measure
Energy efficiency credits scheme	Regulatory	6 Mtoe of energy saved by 2012	Energy service companies, electricity and gas distributors, parties which have taken steps to appoint an energy manager	Existing (to be extended)	January 2005 –
55% tax relief for building refurbishment projects	Financial		End users who own existing buildings	Existing (to be reviewed)	January 2007 – December 2010
Minimum quota of 50% of domestic hot water to be produced using renewable energy sources	Regulatory	% coverage of consumption	End users who own newly-constructed buildings or buildings to be refurbished	Planned	
Tax credit for district heating using geothermal or biomass energy	Financial		End users who connect their properties to district heating networks connected to plants using geothermal or biomass energy	Existing	January 1999 –
Solar photovoltaic feed-in tariff	Financial	3000 MW by 2016 (target currently being updated)	Investors / End users	Existing	August 2005 –
Solar thermal feed-in tariff	Financial	2,000,000 m² of panels installed by 2016	Investors	Existing	May 2008 –
Green Certificates	Regulatory	Feeding electricity from renewable sources into the grid (in 2012, 7.55% of the energy from fossil fuels fed into the grid in the previous year)	Investors	Existing	April 1999 –
All-inclusive tariffs	Financial		Investors / End users	Existing	January 2008 –
Minimum quota for electrical capacity installed using renewable sources	Regulatory		End users who own newly-constructed buildings or buildings to be refurbished.	Planned	January 2011 –
Minimum quota for transport biofuel use	Regulatory	4.5% of transport biofuels fed into the network in 2012	Parties which make fuels available for consumption for automotive purposes	Existing	January 2007 –
Reduction in excise for biofuels	Regulatory		Investors	Existing	1995-2010
Interregional Operational Plan on Energy	Financial	Creation of renewable energy plants, operations to increase energy efficiency	Investors / End users / Public authorities	Existing	June 2007- December 2015

SOURCE: Italian Ministry of Economic Development (2010)

International cooperation mechanisms	Financial	Availability of renewable energy equal to approx. 1.1 Mtoe by 2020	Other countries, Investors, TSOs	Planned	January 2016 –
Further simplification of authorisation procedures	Regulatory		Investors / End users / Public authorities	Planned	2010-2020
Definition of technical specifications (e.g. performance standards for biomass fuels)	Regulatory		Investors / End users	Planned	2010-2020
Support for the creation of district heating and district cooling networks	Regulatory		Manufacturing areas / Residential areas	Planned	2010-2020
Training and informative campaigns	Soft	Changes in behaviour	Operators, designers, regions, local authorities, citizens, companies, etc.	Planned	2010-2020
Support for the development of the electricity network	Regulatory		AEEG#, TSOs, Electricity network operators, Distributors	Planned	2010-2020
Support for the integration of biogas into the natural gas network	Regulatory		Agro-industrial system, gas transmission and distribution network operator	Planned	2010-2020
Sustainability criteria for bioliquids and biomass	Regulatory		Operators	Planned	2010-2020
Kyoto funds	Financial	Creation of renewable energy plants, operations to increase energy efficiency and reduce emissions	Investors / End users / Public authorities	Planned	

4.5 How

How the cases are analysed cover 360° of understanding: initially they were selected and studied in literature, then visited in site and finally reflected with continuous feedback and interaction. The selection was important because each one should respect the fifth SD principles and not often there were enough public information and references. So this cases studies rely heavily on site visits to observe companies and projects in action, meet and interview informants in they "home" environments and "experience" in some how the complex systems. Additionally, key informants were often at ease and eager to show and explain in detail the development of "their" green energy systems (Stake, 1995). The pictures taken during the site visits capture observations and help "to be in the place" with hindsight. The site visits played a crucial role in the research process and the case studies. In order to assure validity and credibility, different sources of information are used, such as interviews, direct observation at the sites, websites, newspaper, articles,… The informants from the companies have also been able to study and comment on drafts of the studies in order to increase the validity. One matter in this kind of research is that important information is often confidential.

At the end those cases were synthesized in homogeneous and structured form putted in the annexes of this manuscript. The common format show the following elements of analysis: localization, historical background, objectives and strategies, activities, focus on environment, focus on job, main actors and promoters, funds and financing, future projects, references and websites. This research adopts a wide perspective on bioenergy, that cuts across several sectors and include a range of actors involved in systems. This broad viewpoint facilitate a greater understanding and designing of the framework. Systems models are mental constructs used by researchers to understand the world and take action (Senge, 1993), clarifying complex relationships of the reality and thus making them open for studies. The representation of each case study with self-done schema helped the researcher to better understand the system, and then to depict findings and concepts in presentations. Graphic schemas highlight the crucial elements that make up green energy systems and their interactions (inputs, outputs, interactions, components, feedbacks). Multiple representations can show the technological aspects, such as conversion technologies and biomass resources, and social aspects, such as policies and actors (Geels, 2004). The systems comprise four main technical aspects, which are biomass, resources, supply systems, conversion technology and energy services. Furthermore, they involved a range of actors, that are connected through a complex and dynamic web of networks, partnerships and policies, like the governance (municipalities), the knowledge network (universities and research centres), the material network (suppliers, farmers,...), the energy network (power equipment manufacturers) and the power companies (energy sellers). In that way it is possible to explore the influence and role of different actors in the implementation of green energy systems. Those networks allow a better understanding of "hidden connection" and ultimately leverage points[4]. Policies and/or actions that apply pressure to leverage points[4] can often be very effective in achieving the desired goals.

Case studies with strong empirical influences can contribute to the research in an interdisciplinarity study in several different ways:

- They can contribute to empirical tests of general ideas and help support or disprove an existing theory by comparing the results to the theories and previous researches.

- By observation of facts and situations not included in existing theory they can contribute to the improve the theory and develop a model.

- Case studies provide an opportunity to systematically examine qualitative data that is not so easy aggregated. Case studies can thus contribute to knowledge about conditions that can only be described in qualitative terms.

---

[4] Leverage points are places in systems where small changes result in large responses or shifts in systems (Capra, 1997; Meaows, 1999).

CROSS ANALYSIS

The **cross analysis of the cases and of the theoretical with the practical parts** gives important results that allow to design the framework, which is continually verified in the Agrindustria best practice. That analysis wants to turn the theoretical knowledge into pathways of change that are suitable to the specific needs and capacities of regions or localities. This demand engaging the relevant parties in all stages of the change process and enhancing their capacity to effectively address the future sustainability challenges. Although satisfying these latter ambitions can and should be guided by the knowledge provided by the methodology, acquiring a thorough understanding of the dynamic and complex systems and their sub-components for which changes are desired become key.

It is useful to think about the relationship between concepts or theory and the research process in terms of **inductive** and **deductive strategies**. **Deductive research** is the process by which a researcher utilises what is known about a particular domain and the **theoretical considerations**, and then describes a **hypothesis to test** through empirical scrutiny (Bryman, 2004). In contrast, this research work is predominantly based on an **inductive strategy** where concepts or explanations are the outcome of research and **observation in the field**. Undertaking innovative research work demands an **interactive process** involving a back and forth between data and concepts, and between formulation and critique (Connell, 1985).

The cross analysis among the selected case-studies shows the benefits of SD that can be evaluated either quantitatively and qualitatively. The quantification is problematic for complex system because it is hard to define flexible boundaries and discriminate the environmental, social and economical benefits by the SD project from other parameters affecting the system. The boundaries define the best uses of energy and material stream, and the analytical framework gives a conceptualisation of systems; in that way the case studies are really comparable.

The development and use of **analytical framework** is such an element, which is connected with the overall research methodology. The analytical framework is a way of conceptualising and interpreting the practical experiences. It is used for identifying, collecting and analysing data, so they can be comparable in vertical and in horizontal way. In this research, the vertical comparison is the comparability of the same parameter among different case studies, the horizontal comparison is the modification of the parameters in the same case-study (*figure 5.0.I*). The analytical framework is helpful especially in the vertical comparison because all the parameters can be considered in parallel; in the horizontal comparison its fundamental the changes in the course of the time, so some case studies do not have an historical background, that can give an evolution of the parameters, like in Italy, and some parameters do not evolve,

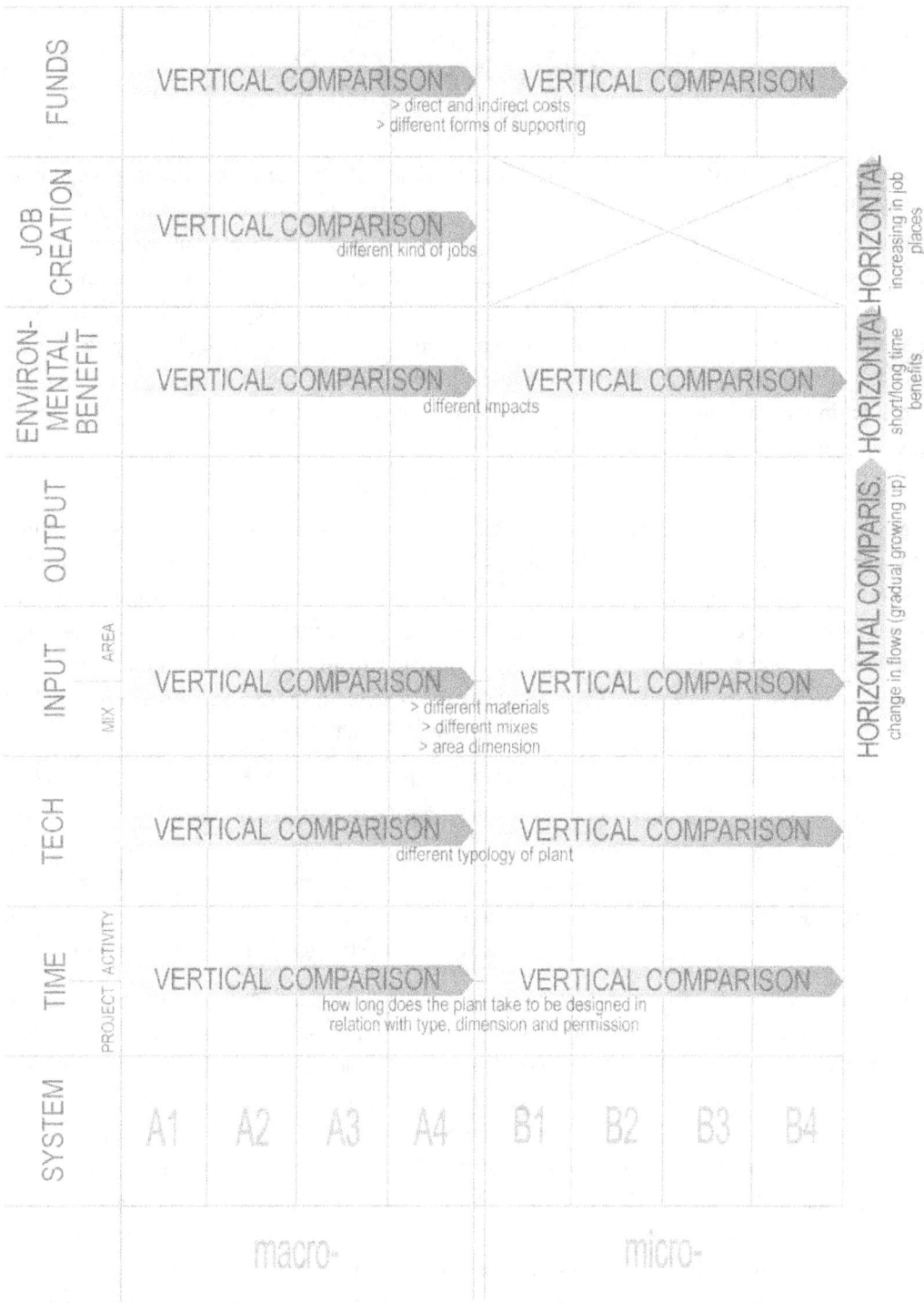

figure 5.0.I: vertical and horizontal comparison of case studies.

like the time spent to design the plant and the time of activity, the technology and the amount of money invested to realize the system.

The analytical format considers:

- the **time** for the project and how long the system is working: those data help to understand the difficulty to design a new power station and if there are some events that can obstacle the success. Furthermore, if the system is working from long time, it is an index of stability and it has also historical data that can show the productivity in the course of time. The vertical comparison allows to understand how long does the plant take to be design in relation with the type, the dimension and the permission. The horizontal comparison does not make sense with that parameter because it is fixed along the time itself.

- the **technology** in the energy plant: the productive systems can be different and more or less tested. The fed materials, with its quantity and quality, and the different kind of energy (electricity, heat, biofuel,...) can give very different technical solutions. The vertical comparison identifies the different typologies of energy technology plants that are actually available in the market. The fixed technology for each case study does not allow a horizontal comparison.

- the **output>input flows of materials** with their quality and quantity, and with the distance of collection: these data are strictly connected with the technology used to transform them into green energy. In macro-systems the quantity and the afferent radius are higher than in micro-systems; furthermore, the type of energy technology can be considerably different. The vertical comparison allows to identify the different typologies of materials and their different mix, and to define the dimension of the areas. The horizontal comparison highlights possible changes in the input mix and if the complex system grew up gradually.

- the **environmental benefits**: in designing bioenergy systems, the territory will be modified with bigger or smaller changes in relation to the dimension (macro- or micro- systems). The vertical comparison reports the different impacts on territories, and the horizontal comparison supports those results with the hint of short and long terms.

- the **job creation**: macro-systems influences the territory in multiple aspects and with the International Labour Organization (ILO) it was studied the parameter that easily can give interesting results in vertical and horizontal comparisons. In that case, the vertical comparison means the different job creation among all the case studies; and the horizontal comparison is the increasing in job places in each case studies. That parameter is not considered for micro-systems because the single plant is so small that does not generate new direct job.

- the **private and public funding** to realize the project: the economical aspects are also considered in this research to verify the sustainability of the systems also at the financial point of view. Those data can give the percentage of the support and its origin (public and private) to understand also the national dynamics of market and policy. The vertical comparison allows to examine direct and indirect costs of different bioenergy systems and how they are supported. This parameter is fixed, so it does not allow a horizontal comparison. It is important to recognise that any analytical format is an attempt to provide some simplistic order to an infinitely complex world. So caution is needed when interpreting results or findings.

The research method for all case studies generally started with observation, then analysis and interpretation, and concluded with comparisons and explanations. The aim of this approach is not to reduce complexity by breaking research problems down into variables that "fit" with existing theory but instead to increase complexity by including the context (Alvesson and Sköldberg, 2000). As with any interdisciplinary research project, this work is related to several research disciplines, that are interrelated and sometimes overlapping.

Utilizing a **combination of *desk* and *field research*** allows a more in-depth understanding of reality from different viewpoints, which is crucial when exploring topics or issues involving a range of actors, like the energy sector. This research is heavily focused on qualitative methods, including literature reviews, case studies, site visits, stakeholder interviews, industry interactions and international agencies connections. The foundation of this research is therefore the use of **diverse kinds of data sources** and a **mix of qualitative methods**. The *desk research* detects the information already written by others, in a important identification of the sources with their reliability. The field research directly observes the reality with empirical and personal experience. It means that the desk research is less expensive and also less original; but it helps to define the headway processes of the research. The combination of desk and field research guarantees a reach and deep understanding of the facts to define an original framework (Celaschi & Deserti, 2007).

Rather than setting out with a proven "one-fits-all" solution in hand to catalyze the development of SD networks, this research primarily focuses on continuously fine-tuning approach so as to make it best suited to dynamic region specific contexts.

5.1 Macro-systems comparison

The cross analysis among the macro-systems gives interesting results about the project and the activity time, the different technologies, the output>input flows of materials, the distance of material collections, the environmental and the social benefits, and the private and public funding (*figure 5.1.1*).

Time

In Sweden, these kinds of projects started in the 80s with a process of long participatory planning lasting about one decade. Those macro-systems involve both public and private sectors,

CROSS ANALYSIS

SYSTEM	TIME			TECH	INPUT		OUTPUT	ENVIRON-MENTAL BENEFIT	JOB CREATION	FUNDS	
	PROJECT	ACTIVITY			MIX	AREA				INVESTMENT	FUND
LINKÖPING	7 y (89-96)	14 y (2010)		anaerobic fermentation	66% ABATTOIR / 15% OTHER MATERIALS / 10% INDUSTRIAL ORGANIC WASTE / 9% PIGS & CATTLE MANURE	50 km	biogas 7.7 million m3/y; bio-fertilizer	-) reduction of waste -) reduction of GHG emission -) reduction of noise and pollution in sensitive areas	-) participative design -) spread of know-how -) 1,000 employees in Tekniska Verken (30 direct job + 100 indirect job) -) 35 direct employees in ENA Energy	8.7 M€	1.7 M€ (19.5%)
ENKÖPING	13 y (81-94)	16 y (2010)		steam boiler (CHP)	50% WOOD CHIPS / 20% SALIX / 15% SAWDUST / 15% PEAT	70 km	heat 55 MW/h; electricity; bio-fertilizer; terracing material	-) reduction of waste -) reduction of GHG emission -) non-chemical fertilization -) reduction of nitrogen and phosphorous		35.6 M€	5.3 M€ (14.8%)
VÄXJÖ	17 y (80-97)	13 y (2010)		steam boiler (CHP)	70% WOOD / 15% PEAT / 7% SAWDUST / 8% BARK	120 km	heat 66 MW/h; electricity; bio-fertilizer	-) reduction of waste -) reduction of GHG emission -) non-chemical fertilization -) reduction of nitrogen and phosphorous -) reduction of energy consumption -) FSC forests	-) wide variety of skills -) 20 direct employees -) 8,000 indirect jobs	47.5 M€	11 M€ (23%)
VIGNOLO	3 y (89)	1 y (2006-2009)		anaerobic fermentation	84% COW LIQUID MANURE / 9% OTHER / 7% CHICKEN MANURE	40 km	biogas 2.4 million m3/y; electricity; bio-fertilizer	-) reduction of waste -) reduction of GHG emission -) non-chemical fertilization	-) 4 direct employees in Vignolo plant -) 2 direct employees in R&D -) indirect jobs as truck, mechanical, selling operators	7.5 M€	private loan

figure 5.1.l: macro-systems comparison.

and the sharing of problems and solution improve the satisfaction of all participants. The commitment of different people is important in order to realize the proposed measures and start cooperation. Sweden is a small country with around 9 million inhabitants and everyone knows everyone else in a given subject area. This facilitates the dialogue between the authorities, companies and researchers. The country is densely forested, has an active forestry that endeavours to find a balance between the environment and production, and where bioenergy is integrated with both forestry and the extensive forest industry. Sweden has a high heat requirement that, in combination with extensive district heating system, provide excellent base for effective Combined Heat and Power (CHP) generation from solid biomass. Even if the participative design takes long time to reach the goals, because many meetings and many compromises should be taken among the parts; it has more chance to be successful. The favourable outcomes are confirmed also by the long lasting of the systems, they started in middle 90s and now they are still working and improving the area. They are dynamically stable and have historical data with the productivity in the course of time.

In Italy, the situation is quite different, because it is a totally private project without a participative design and it started around two decades later than the Sweden case studies. The system design in Vignolo (CN) is made by a single firm (Marcopolo Environmental Group) that in 3 years define the complex system and ask for the permissions to the public parties, but without involving them in decisional aspects. The shorter time in designing, help to catch up some delay in comparison to the activities in Sweden, where those kind of systems are working since around one decade and half. The Vignolo systems started the tests in 2009 and it is fully operational in 2010.

Technologies

In macro-systems two different kind of technologies are used to process the biomass: the anaerobic fermentation (Linköping and Vignolo) and the steam boiler (Enköping and Växjö).

The anaerobic fermentation is a series of processes in which microorganisms break down biodegradable material in the absence of oxygen to release biogas and bio-fertilizer. The digestion process begins bacterial hydrolysis of the input materials in order to break down insoluble organic polymers such as carbohydrates and make them available for other bacteria. Acidogenic bacteria then converts the sugars and amino acids into carbon dioxide, hydrogen, ammonia, and organic acids. Finally, methanogens convert these products to a mix of methane and carbon dioxide (biogas). Furthermore, the digestate from this process is very nutrient-rich and can be used as soil fertilizer.

The steam boiler system is used to simultaneously generate both electricity and heat (cogeneration), so it is one kind of the CHP system. A steam turbine is a mechanical device that extracts thermal energy from pressurized steam, and converts it into rotary motion. It is suited to drive electrical generator, furthermore, by capturing the excess heat, it potentially reaches efficiency up to about 90%. CHP is most efficient when the heat can be used on site or very close to it. Cogeneration plants are both used for district heating systems of the cities. Furthermore, the ashes from this process are very nutrient-rich and can be used as soil fertilizer, in Enköping the bottom ash is mixed with sewage sludge to fertilizer the salix crops, then used as biomass for the same CHP. Finally, the ashes can be also used as terracing material.

Output>input

The anaerobic fermentations have biomass with high moisture, in Linköping it comes mainly from the municipal abattoir (organs, carcasses, blood, and hides), in Vignolo from the local farms (cow and chicken manure). From those different materials, collected in an area of 50 km for the former plant, and of 40 km for the second one, they produce respectively 7.7 M m^3/y and 2.4 M m^3/y of biogas.

The steam boilers use wooden biomass, mainly wood chips. Those materials are collected in larger areas than the input for the anaerobic systems: in Enköping, they come from an area of 70 km and produce 55 MW/h of heat; in Växjö, they come from an area of 120 km and produce 66 MW/h of heat. Furthermore, the horizontal comparison of these two case-studies shows how the CHP systems grew up gradually with a continuous modification of previous power plants.

Environmental benefits

The main environmental benefit, that is consistent in all case-studies, is the reduction of waste, because in these systems, the complex and interdependent relationships among different system components enable the use of waste from one process as input to another. The dynamics of nutrient flows in natural ecosystems can solve the pressing problems of actual industrial productions.

Another benefit shown by all case studies is the reduction of GHG emission; this data is complex to obtain because of many different elements should be included in the analyses. The GHG emission reductions have been calculated by experts of the respective case studies. The results are considerable at the environmental level, because the GHG in atmosphere absorb and emit radiation within the thermal infrared range, causing the greenhouse effect.

The third benefit in common with all case studies is the use of non-chemical fertilizer, because from both the anaerobic fermentation and the steam boiler process is obtained a biofertilizer, respectively digester and ash. Their nutrients are rich in nitrogen and phosphorus to stimulate plant growth and to build soil organic matter. Biofertilizers can be expected to reduce the use of the chemical fertilizers and pesticides.

In the CHP case studies, there is another important benefit for the water systems: the reduction of nitrogen and phosphorous in the local rivers and lakes (the Mälaren in Enköping, and the Mörrum in Växjö). Those substances increase the phytoplankton in the water body, with many negative environmental effects (i.e. anoxia). In that way the rivers are safe and secure for recreation, fishing, and aesthetic enjoyment.

In the case study of Linköping, an important environmental benefit is underlined: the reduction of noise in sensitive area. The use of biogas as a clean fuel gives answers to current smell and noise annoyances in traffic congested areas, with also social benefits for drivers and passenger.

In the case study of Växjö, two environmental benefits are considered: the reduction of energy consumption and the forest management in accordance with the Forest Stewardship Council (FSC). One of the Växjö Municipality's goals is the reduction of electrical energy consumption by at least 20% per inhabitant, until 2015. To reach this ambitious goal the power company of the city is working to real projects, like the energy efficient timber building that will help the citizens to reduce their energy consumption by 5% in three stages. Furthermore, the Växjö Municipality will increase the productive forestry land managed in accordance with FSC from 47% in 2005 to 60% in 2015. The FSC has 10 principles and associate criteria that form the basic all FSC forest management standards[1].

Job creation

The human dimensions are important to understand the factors that influence decision makers and the diffuse wellness. This aspect so tricky was developed with the ILO because it is the specialized agency of the United Nations deals with labour issues. The method used to evaluated the social aspects and in particular the job creation guarantees results that are useful for this research and for their labour statistics. The vertical comparison shows how many direct jobs are created in the different case studies: Linköping has the higher number of employees in Tekniska Verken AB (1,000), then 35 in Enköping, 20 in Växjö, and at the end, just 6 in the newest Vignolo. These data show the great opportunities for green jobs[2], without considering the indirect jobs connected with those complex systems, like truck, mechanical, and selling operators. Furthermore, the range of profiles stretches from highly skilled research and development or management functions through to technical and skilled levels to the relatively low skilled. The largest numbers of these green jobs is concentrated in the bioenergy sector and the

---

1   The process for developing FSC standards are: transparency, independency, and participatory.

2   The ILO reports define "green jobs" as work in agriculture, industry, services and administration that contributes to preserving or restoring the quality of the environment. They help to cut the consumption of energy, raw material and water through high efficiency strategies, to de-carbonize the economy and reduce the GHG emissions, to minimize or avoid altogether all form of waste and pollution, to protect and restore ecosystems and biodiversity. The notion of a green job is not absolute, but there are shades of green and the notion will evolve over time.

recovery of raw materials. The contribution that green jobs makes to clean economy growth, development and poverty reduction depends on the quality of these jobs (Committee on Employment and Social Policy, 2008). The horizontal comparison shows how in some case studies the new job places had a gradual increase. In Enköping, for example, the Enköping Värmeverk AB had just 4 employees in the 70s, then 14 in the early 80, 20 in the late 80, and 35 nowadays.

Private and public funding

The economical aspects show how all the selected case-studies are sustainable also at the financial level. In Swedish cases, the funds came from public financings, instead of in Italian case, the fund came from a private bank[3]. For the anaerobic fermentation plants the investment costs were about 8 M€; the steam boilers had different costs for their different dimension. That one in Enköping produces 55 MW/h of heat and cost 35.6 M€, that one in Växjö produces 66 MW/h of heat and cost 47.5 M€. The last one had also the higher percentage of funds by the EU (23%), with a total of 11 M€. The Linköping biogas plant was funded with the 19.5% of the total investment by the government, with a total of 1.7 M€. At the end, the steam boiler in Enköping was funded with the 14.8%, that means 5.3 M€.

5.2 Micro-systems comparison

The cross analysis among the micro-systems gives corresponding results as among macro-systems, even if the dimensions considerably change. The social benefits are qualitatively described in each case studies[5]. A quantitative evaluation of, for example, job creation, was not possible due to the small size of the projects (*figure 5.2.I*).

Time

For smaller projects than the macro-systems, the design time is really shorter: at least one year, in Sweden. On the contrary, the Italian case study does not correspond with the previous data, because:
- it is a totally new project for the country;
- the investment is relatively high with the firm's budget;
- the procedures for permission are longer than in Sweden.

Considering the last point, it is also important to notice that all micro-systems have private part as main promoter, that involves the public part just for building and emission permissions.

---

[3] The percentage and the amount of the bank financing were given as confidential information, so they will not be published in this thesis work.

In all case studies, the design time started between 2003 and 2006; so in Sweden, the plants started the production activities at the latest in 2007, but in Italy just in late 2010.

Technologies

Each micro-system has a different technology, because there are many technical solutions available for small productions, that are very flexible and suitable to different needs.

In Bioagro system, the waste from agricultural fields is pelletized and then burned. Sawdust, straw, hemp, and bark are mixed in a different percentage[4], and then pushed in a pellet mill, conditioned and thermal treated, and, at the end, cooled down. This process gives a homogeneous product of high quality to be burned in stoves and boilers, offering combustion efficiency of over 90%. The ash from this process can be used as soil fertilizer, avoiding the chemical one. Emissions such as NOx, SOx, and volatile organic compounds from pellet burning equipment are very low in comparison to other forms of producing heat from combustion.

In Villa Ödman system, the household wastewater[5] is recovered by struvite precipitation and zeolite absorption techniques. Bertil Eriksson designed a comprehensive system treats all water born waste from kitchen, shower, and toilet through a combination of ventilation, heat recovery, water purification and drainage systems. This integrated system is covered by a series of patents, which form the backbone for the SplitBox technology. Furthermore, the digestate from this process is very nutrient-rich and is used in the private garden as fertilizer. The maximal recovery and reuse of spill-energy, from the wastewater and the ventilation air used in that complex system, in combination with good isolation techniques, result in about 50% of energy savings. Consequently, this contributes to 50% of GHG reduction, in comparison to fossil fuel energy source.

In Pugerup farm, the wasted straw from agricultural field is burned in a round bale burner to heat air, which is in turn used to dry the grain, and to heat water, which is in turn used to heat the buildings. The flue pipes pass the entire length of the water jacket before exiting the furnace. This helps to recover heat that would have otherwise been lost in exhaust gasses. When the water temperature drops below the set temperature, the fan comes on. The air is usually distributed evenly across the chamber through a manifold. Forced air greatly improves the rate of combustion and therefore the rate of heat production by the fuel. The steam boiler is handy and requires a minimum of service and maintenance. It is a highly efficient straw boiler with good combustion and an efficiency exceeding 88%.

---

4   The percentage of the materials were given as confidential information, so they will not be published in this thesis work.

5   The water from the bathroom amounts to 350 kg/day; the organic material and the water from the kitchen amount to 140 kg/day; and the urine, faeces and paper from the toilet amount to 58 kg/day.

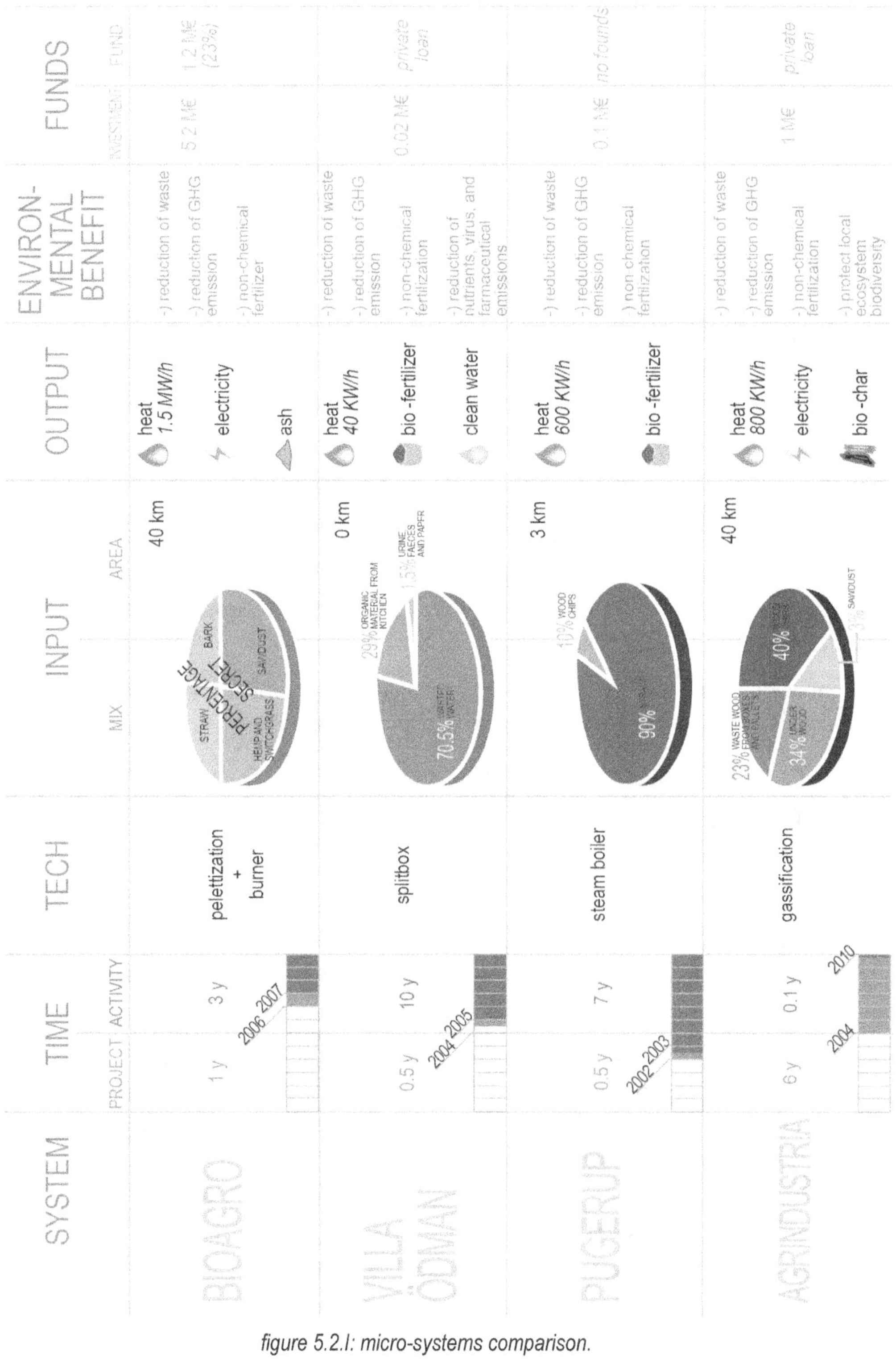

CROSS ANALYSIS

SYSTEM	TIME		TECH	INPUT		OUTPUT	ENVIRON-MENTAL BENEFIT	FUNDS	
	PROJECT	ACTIVITY		MIX	AREA			INVESTMENT	FUND
BIOAGRO	1 y	3 y 2006-2007	pelettization + burner	STRAW, HEMP AND SWITCHGRASS, PERCENTAGE SECRET, BARK, SAWDUST	40 km	heat 1.5 MW/h; electricity; ash	-) reduction of waste -) reduction of GHG emission -) non-chemical fertilizer	5.2 M€	1.2 M€ (23%)
VILLA ÖDMAN	0.5 y	10 y 2004-2005	splitbox	28% ORGANIC MATERIAL FROM KITCHEN, 70.5% WASTED WATER, 1.5% URINE FAECES AND PAPER	0 km	heat 40 KW/h; bio-fertilizer; clean water	-) reduction of waste -) reduction of GHG emission -) non-chemical fertilization -) reduction of nutrients, virus, and farmaceutical emissions	0.02 M€	private loan
PUGERUP	0.5 y	7 y 2002-2003	steam boiler	10% WOOD CHIPS, 90% STRAW	3 km	heat 600 KW/h; bio-fertilizer	-) reduction of waste -) reduction of GHG emission -) non chemical fertilization	0.1 M€	no founds
AGRINDUSTRIA	6 y	0.1 y 2004 / 2010	gassification	40% WOOD, 23% WASTE WOOD FROM BOXES AND PALLETS, 34% WOOD NUTS, 3% SAWDUST	40 km	heat 800 KW/h; electricity; bio-char	-) reduction of waste -) reduction of GHG emission -) non-chemical fertilization -) protect local ecosystem biodiversity	1 M€	private loan

figure 5.2.l: micro-systems comparison.

In Agrindustria, the wasted biomass from the company and from the surrounding activities is processed with gasification. That process converts woodchips, under-wood and other wooden materials into carbon monoxide and hydrogen by reacting the raw material at high temperatures with a controlled amount of oxygen. The resulting gas mixture is called syngas and is itself a fuel used in a gas engine to produce heat and electricity. The pyrolysis process occurs as the carbonaceous particle heats up: volatiles are released and bio-char is produced. Bio-char is of increasing interest because of concerns about climate change caused by GHG: it is a way for carbon to be drawn from the atmosphere and is a solution to reducing the global impact of farming (and in reducing the impact from all agricultural waste). The system realizes an overall benefit through reduced fossil fuel consumption at the company's plants, which is reflected in increased net energy ratios.

Output>input

The micro-systems have different technologies because the input materials and the specific environmental contexts are quite different. However, in each case there is a production of heat and of materials used as fertilizer; furthermore, in some cases also of electricity (Bioagro and Agrindustria). The higher amount of heat is produced in Bioagro system, and the lower one in Villa Ödman for its internal consumption.

The input materials are quite different, but mainly dried biomass (sawdust, straw, wood chips,...), only in Villa Ödman it is wet, with high percentage of domestic waste water.

Environmental benefits

Three environmental benefits are constant in all case studies: the reduction of waste and of GHG emissions, and the use of non-chemical fertilizer.

The reduction of waste is guarantee by the complex and interdependent relationships among different materials that from waste becomes input in other processes. The dynamics of nutrient flows in natural ecosystems solve the problems of domestic and micro-industrial production.

The reduction of GHG emission is evaluated with complex systems that include many variables and parameters. This research considered these data given by the experts that analysed each case-study because the GHG concentration in the atmosphere modify the global clime, and because is caused mainly by the growing use of energy.

The use of non-chemical fertilizer is important for the healthiness of soil, and the farmed products, as well as water bodies. Their nutrients are rich in nitrogen and phosphorus that stimulate the plant growth, build soil organic matter, and provide wild life benefits. Besides the use of non-chemical fertilizers, the use of water in maintaining the health of the soil is also important.

Furthermore, in Villa Ödman, a reduction of nutrients, virus, bacteria and pharmaceutical emissions is evaluated. The nutrients, especially potassium, extracted from urine through a combined precipitation and absorption process followed by an oxidation by an oxidation of wastewater leaves pure water behind. The dry, bacteria- and virus- free substance is used as fertilizer.

With the system in Agrindustria, the local eco-system is also protected and the biodiversity preserved. This particular case involved the mountain communities in cleaning out the underbrush, and the local farmers in farming mixed associated woodland using the biodynamic method. Many other benefits are related to its activities, like improving the attention for biodiversity protection, maximizing plants productivity, and optimizing storage plants functions.

Private and public funding

The economical aspects in the micro-systems are often confidential, because the private funding are used.

That is the case of Villa Ödman's landlord, who spent 20,000.00 € for the SplitBox, but the whole house building should be taken in consideration[6] in the evaluation.

Pugerup boiler had a relatively small investment of 0.1 M€, so the owner did not ask for loan.

Agrindustria gasification plant cost 1 M€ and it was totally carried by the company, with the help of a funding of a private local bank[7].

The only one that was supported by EU funds was Bioagro, because it won a Life Programme grant in 2006 and lasted 36 months. The total investment was 5.2 M€ with a funding by EU of 1.2 M€ (23%). It co-finances innovative pilot and demonstration project supporting environmental and nature conservation projects.

5.3 Theory and practice cross analysis

Systems thinking can provide a clear vision for considering the complexity of factors adequately, while understanding how they are connected. Experts focusing on their own disciplines may act within a narrow scope without considering the linkages of their field with other disciplines, leading to a "fragmented" policy agenda (Wjikman, 2008). An open system that integrates ideas from stakeholders from all different levels and fields could bring about creative policy solutions that would not be possible with thinking that is strictly linear. While the sectoral approach

---

[6] The entire amount and any loan are confidential information.

[7] The percentage and the amount of the bank financing were given as confidential information, so they will not be published in this thesis work.

encourages public-private partnerships, it may be difficult to involve those who do not identify themselves with the specific sector, and neglect major emitters and energy users. This view may lead to delayed or duplicated efforts causing a larger expenditure of resources.

To make objective observations and to be neutral, the logical results are based on empirical findings. Case study approach is chosen because it is well suited to study complex and contemporary phenomena in their real life contexts. In this continuous process of research and learning, also the theory part is improved by critical reflection. This is one of the most impor-tant stage of this research as this is where most of the analyses are carried out, the learning is made explicit and new actions for improvement are formulated in light of the analysis. The SD principles are reformulated throughout the process based on the specific changes in the focal context and in bioenergy systems. Although they are a contribution to practise, they are an explicit goal of the theory research to generate scientifically valuable knowledge (*figure 5.3.I*).

1) **Output>Input:** *as in nature what is not used by a system becomes a raw material for the development and survival of someone/something else, in the production process the waste (output) of a system becomes an opportunity (input) for another one, creating new economic opportunities and new jobs.*

In that sense renewable (wood, biomass, waste) or perpetual (sun, wind, water) resources to produce green energy can be fine. In particular it is a very good point starting from what usually is considered waste, so the benefit can be higher. It also becomes a matter of introducing and adding flexible energy technologies and designing integrated energy system solutions. Not only technological changes are required in order to generate further sustainable development.

2) **Relationship:** *it is important to consider, more broadly, all the networks of components that made the food system, including materials (resources) and energy, which are used, captured and stored through different stages of the product life cycle. Understanding the pattern of materials and energy flow and investigating where it can be improved can allow us to find entry-points for designing more sustainable food system.*

Availability of a local energy system can develop a territory, so an energy system so integrated with the area can straighten the relations with different actors, from institution, to private sector and communities. A participated design helps to create awareness and heightens the level of commitment on the part of each co-designers. Involving the private sector can encourage direct dialogue with sub-national authorities and open new markets that help identify new financial sources for sustainability projects. Governments should encourage active investment from the private sector, especially for eco-innovations and the transfer of green technologies. Such investments can be presented as win-win-win relationships, which, in addition to improving the environmental sustainability, allow companies to be socially responsible and establish their position in a community. They could also bring about economic gains for the region and the people in the community.

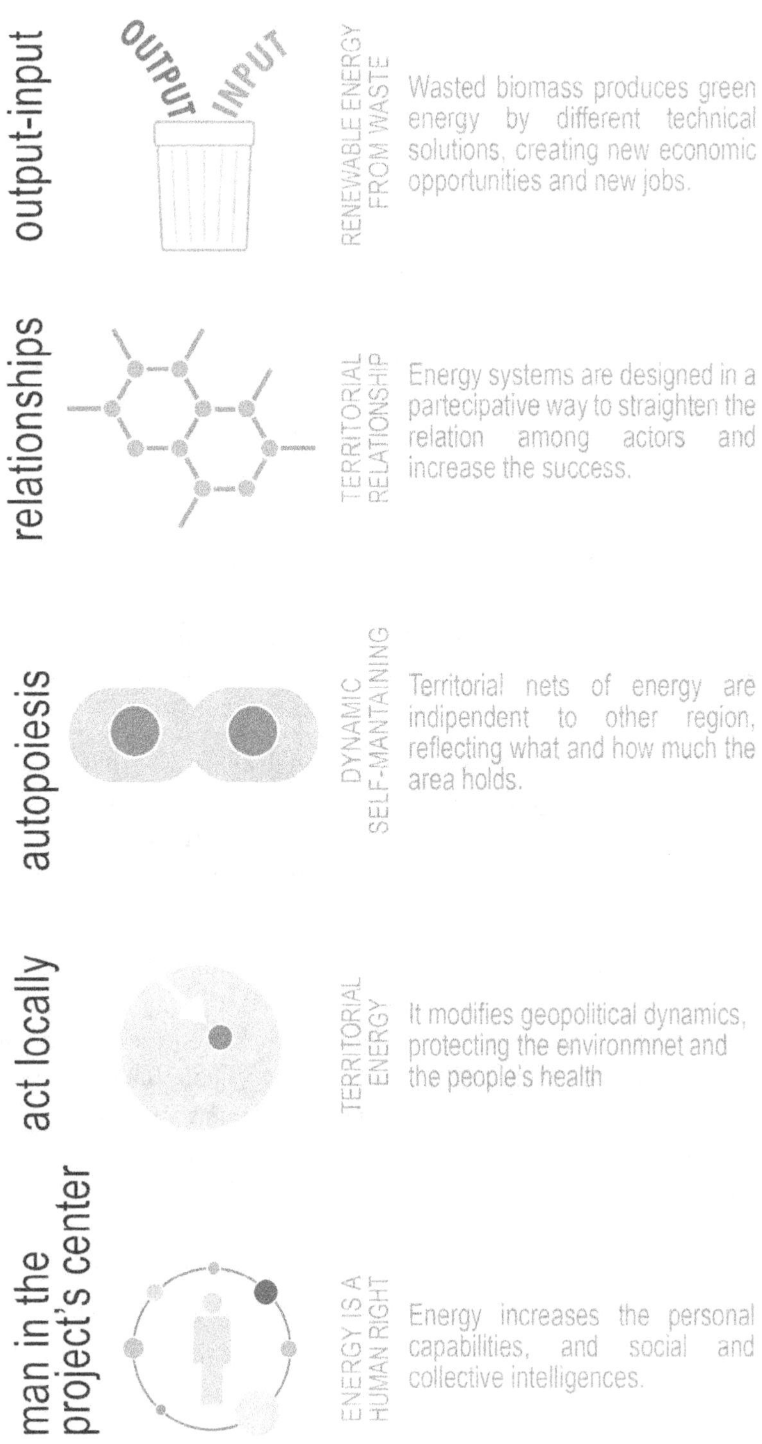

CROSS ANALYSIS

output-input — RENEWABLE ENERGY FROM WASTE — Wasted biomass produces green energy by different technical solutions, creating new economic opportunities and new jobs.

relationships — TERRITORIAL RELATIONSHIP — Energy systems are designed in a partecipative way to straighten the relation among actors and increase the success.

autopoiesis — DYNAMIC SELF-MANTAINING — Territorial nets of energy are indipendent to other region, reflecting what and how much the area holds.

act locally — TERRITORIAL ENERGY — It modifies geopolitical dynamics, protecting the environmnet and the people's health

man in the project's center — ENERGY IS A HUMAN RIGHT — Energy increases the personal capabilities, and social and collective intelligences.

figure 5.3.I: SD principles applied to bioenergy field.

3) **Towards autopoiesis**: *in nature self-maintaining systems sustain themselves by reproducing automatically, thus allowing them to define their own paths of action. In this way the system is naturally led to balance and preserve its independence. If also in the food system we would start in terms of autopoiesis, it could be possible to allocate efficiently and distribute equally the material and energy flow.*

A territorial net of energy is independent from other countries or regions, because it is simply based on what and how much the area has is both strong and flexible. That system can easily and fast change if the initial territorial condition change.

4) **Act locally**: *as an eco-system is deeply influenced and shaped by its habitat, the same happens for any other type of system. Based on the opportunities provided by the local context, new opportunities can be created by reducing the problems of adaptability due to "general" solution and increasing people's participation.*

5) **Man and his community at the centre of the project:** *the product has become fulcrum of a paradigm of values and actions, as the economical wellness, the quantity of currency resources, the wish of belonging to a social status, that shape negatively consumption choices. The systemic approach, instead, questions the present industrial setting and proposes a new paradigm where at the centre of each productive process there are social, cultural, ethical and biological values that are every man shares.*

The rights of the citizens are callable on energy ground to enforce a proficiencies growth, in a process of individual capabilities. The citizens consume energy, but they give it back in form of social and collective intelligence. Supporting programs of interdisciplinary studies and SD would help train professionals to adopt a systems approach in thinking about sustainability. Governments should engage these professionals in the policymaking process.

The government consider energy as a factor for achieving competitiveness and should guarantee it to the entire society as common resource. It takes on a supranational role, creating worldwide political and economic dynamics. The countries are (inter)dependent one to the other and those relations can change by the decision of few people, but can have consequences on the population. For citizens, energy is a costumer's right and should be safe for the environment and for the people's health; for (regional and national) administration, energy is a need and moves complex geopolitical strategies; for companies, energy is a profit and contributes the economical growth.

PART III

GOAL

DESIGN AND TEST FRAMEWORK TO LOCAL ECONOMIC DEVELOPMENT BY SYSTEMIC DESIGN

With the experience and expertise gathered, the **SD becomes a minded methodology** to design energy systems that use regional sources and create employment. Knowledge and experiences are used to give direction rather than conclusive answers, modelling the framework to LED and describing the consequences of changes.

This research supports that SD is a tool with the capability in contributing to the changes necessary to move towards a more sustainable energy system. This potential is evidenced by the resource efficiency gains by the operational relations and the networks that transform output in input. In the real experiences, there is sufficient justification to the validity of the assumption that many industrial possibilities for additional synergistic relations exist, but remain unexploited. Therefore, rather than waiting for the self-organized development of hidden connections, catalysing their development through **conscious SD interventions** can be seen as a **proactive approach to addressing sustainability challenges**. For this reason, the designed framework is effective in studying and planning dynamic local bioenergy systems.

Each principles should be taken in consideration during the design and action phases, because all dimensions of the sustainability are covered. The first SD principle about the **centrality of human beings** in systemic project gives a philosophical placement. Thus one of the prime considerations is the importance of people in taking decisions and making the SD happen. Consequently, in regards to the efforts aiming to catalyse the development of SD networks, a conclusion calls for placing the main emphasis on the **human and organizational dimension in SD developments**, and not on technical one. Although some levels of knowledge about region-specific technical aspects are useful to learn from and build in the territory, satisfying the local needs with local capacities. This calls for investigating areas related to local natural and human resource endowments, local consumptive processes, and logistical, managerial, and knowledge capacities of private enterprises. The fundamental and active principle for SD is the **transformation of output in input**: if the theoretical and the practical actions don't start from that point, would be not possible to reduce (or avoid) the emissions into the environment and create the **relationships** named in the third principle (relationships). The relations are so important that can fail the entire project. At the end, l**ocal actions** of project and **its autogeneration** are strictly connected because only a system designed in and for a specific context can live through fast changes.

This research provides **empirical evidence to support these principles** and to give the right importance of addressing the human activities in sustainable developments. The element of this methodology that facilitates the use of the learning gained through conscious reflections on real case-studies is useful to guide the planning of the framework that was tested in an Italian micro-case study. Its usefulness is linked to the fact that efforts make system real, and mistake are often as important as the successful trials. The proper test on a real case applies the framework to its development. Knowledge that is critical for this research and for the enhancement of SD theory requires initiating and taking part in the change process together with relevant decision maker. This research improves the practices of coordination functions and the theoretical approach enabled a significantly more thorough understanding. The **theoretical model was tested and improved with the practical experimentation in Agrindustria**, so the designed framework is already verified and the pilot project is replicable (*figure 6.0.I*).

The micro-firm finds the material to be used to produce energy for its own manufacture, within 40 km, selling surplus and activating a number of positive economic outcomes, due to an optimized exploitation of outputs as inputs of other production systems. The different new products put on the market in highly profitable sectors (i.e. cosmetics, pets and mechanic products) they all stem from the clever use of waste deriving from different manufacturing processes. The enhancement and exploitation of resources based upon their quality features has fostered the creation of new products and services boosting the economic growth and development, both within the firms and its own operational context, sparking collaborations aiming to develop in an integrative with the territory. Several types of biomass are recovered from the surrounding area which are currently unrecognized as valuable, such as the derivatives of local products (poplar bark, wood chips, scraps from the production of crates and pallets) and by cleaning out the woods of mountain communities. The company stands on a strategic spot that allows it to obtain virgin wood by cleaning out the woods of its territory. This business allows many families to remain in the valley because it is located at the natural bottom of the valleys of *Gesso-Vermenagna-Pesio*, *Stura*, *Grana* and *Maira*. The project for a micro-power plant for Agrindustria specifically entails conversion by gassification of biomass such as chipped virgin wood with the addition of flour from nutshells and corn cobs derived from the in-house production of granules. Though this plant is small it adopts innovative technology and has an important feature: it is easy to operate and maintain.

The majority of the framework reported below is based on data gathered through the **direct experiences** of case studies and the **direct participation in coordination efforts** trying to catalyse the development of the complex system in Agrindustria. These include data drawn from first-hand observation material, diary notes, meeting recordings and minutes, interview transcripts, and oral and written correspondences. This trial of data is supplemented by **personal recollections of discussions at key-decision momen**ts. Formal and informal communications with all actors constituted the major source of information.

SD moves towards becoming a tool, which facilitates systemic changes. This way will make a

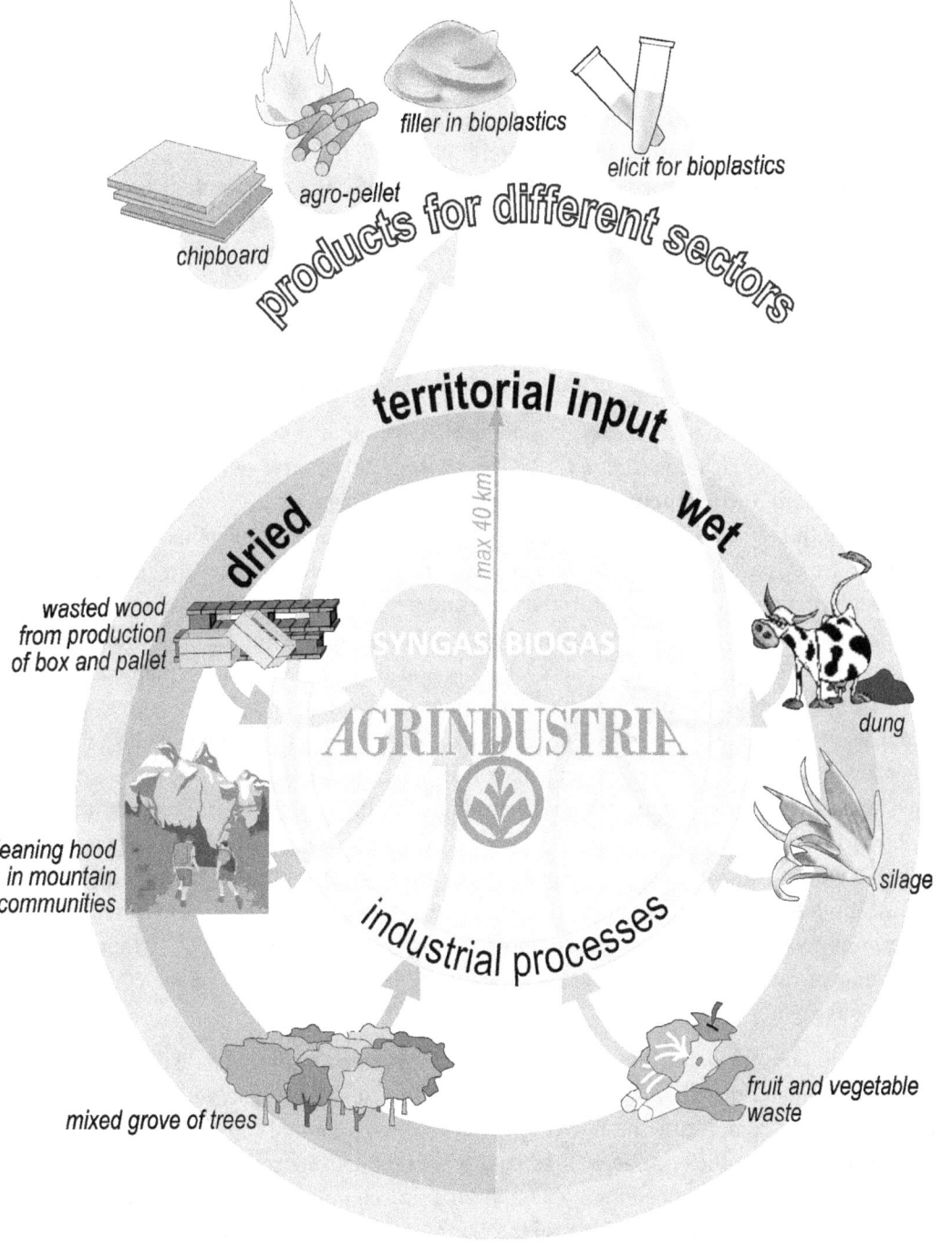

figure 6.0.I: tested framework in Agrindustria with material and energy flows.

more solid contribution to sustainable efforts. It is also dependent on renewed industry-society interactions, particularly with local communities.

The framework advantage is that it can be applied for many different purposes and in different environments while still mantaining the theoretical structure of the relationships to be investigated (Roos, Graham, Hektor & Rakos, 1999). The **framework helps to identify critical factors** for energy territorial nets and **promotes the wise and efficient use of public and private funds** for bioenergy development.

6.1 Territorial nets of energy for Local Economic Development

The framework develops distributed energy systems, that are small-scale, site-appropriate, resilient power generation facilities (*figure 6.1.I*). The **new model of energy production with sustainable and agile infrastructures can develop a region.** Bioenergy produced in **small plants** and **distributed in the territory** helps the success and the sustainability. Such agile system can be a new paradigm for both **energy efficiency** and **reliability** for any region or country. The adoption of more systemic productions and person-cantered approaches, including more innovative capture and generation of energy holds great potential to create LED. When the linkages between materials, energy, people and their knowledge are mapped out clearly, efficient pathways towards sustainable ways to use and re-use untapped resources will become apparent. Using these networks, energy systems can be collectively innovated, so they are beneficial for the people the environment, and pleasurable for all. The wasted biomass in a boudaried area can be processed to obtain new products for different industrial sectors, after this nobilitazation, the further waste can be used to produce bioenergy with the proper technology. In that way the area becomes **self-sufficient at energy point of view, increasing the businesses, the job places** and **reducing the waste** and **the emissions** in the environment. The area around can use exactly the same method, but the result will be different. In that way the regions appear as a patchwork of small sustainable units that can easily change when the initial conditions are not the same.

Massive utilization of renewable and distributed resources tends to blur the distinction between transmission and distribution, and to accentuate the complexity and volatility of grid operation. Increasing complexity, the power grids grow their reliability in demand and requirement; security and efficiency as well as environmental and energy sustainability concerns continue to highlight the need for a leap in harnessing communication and information technologies. These technologies can help also the planning with an **interactive interface** that **can show the relations instantaneously**. The framework facilitates convergence of needs and implementation of necessary analytical capabilities (Moslehi & Kumar, 2010). The prospect of having an increasingly large number of generators connected to various points of the grid will

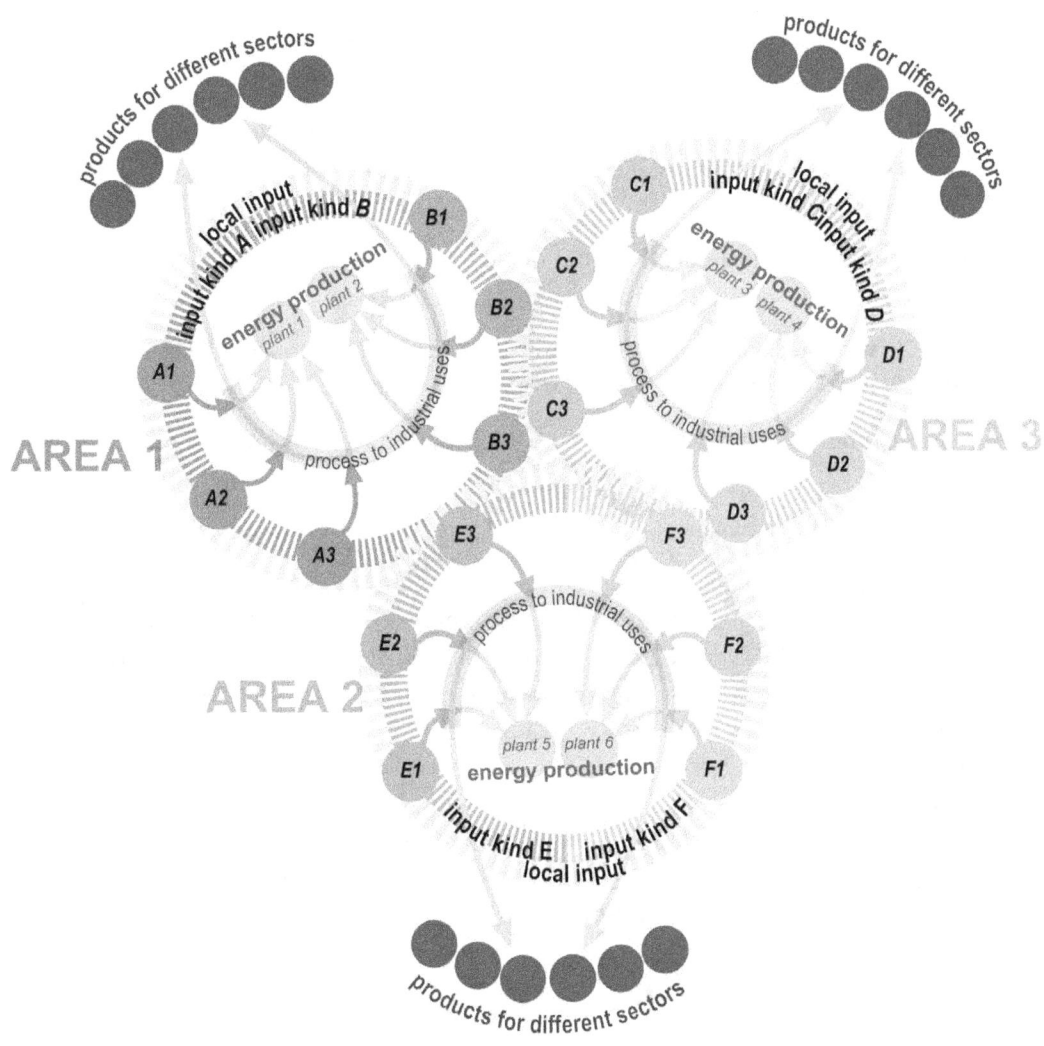

figure 6.1.l: bioenergy framework.

increase the complexity of the current systems and make assessments of technical feasibility (and economics) more difficult. Particularly, grids will become more active in nature and less predictable in behaviour.

Renewable resources generally have adverse impact on grid reliability due to the following factors:

- **load flows**: grids have commonly been designed on the principle that load flows occur in one direction and generally reduce along the length of each distributor;

- **reactive power flows**: some bioenergy technologies are induction machines with no steady-state reactive power generation capability, this can reduce, rather than increase, available capacity in the upstream grid;

- **voltage regulation and fault levels**: in traditional grids fault levels are controlled by network impedance, injection by generators can lead to critical fault levels;

- **power quality**: systems using inversion can create harmonic distortions in the form of the supply voltage, a problem that can be ameliorated by electronic power converters;

- **transient instability**: need to re-design network protection systems due to embedded generation;

- **systems security**: the challenge of maintaining a fairly constant system frequency under high penetration rates.

The role of SD is evaluated in the context of regional developments towards distributed economies. These economis aim to increase the share of thigtly networked, small-scale, flexible energy production units that are more attentive to local resource endowments and to local needs; and that prioritize social and environmental quality aspects of their products and processes. Such systems appear to be better positioned to address the environmental and social dimensions od sustainability at the territorial levels. For example, the REI's Corporate Social Responsibility Program Manager, Kevin Hagen confirms that their choice to buy green energy gives to the company the security to be safe from the increasing of the conventional energy cost and reduced the GHG emissions (Goleman, 2009).

The development of local systems that are alligned with the principles of SD and are supported by the relations among regional systems demands concurrent transitions in various economical, technical, legislative, and social domains. The development of SD systems involves regional actors, linked to technical, informational, political, economic, and human considerations, affecting the caracteristic of regional economy and organizational behaviour. For sure, the presence of economically viable and technically compatible complementaries is an important prerequisite for the development of such systems.

The regional sources to produce energy are usually renewable (wood, biomass, waste) or perpetual (sun, wind, water), unless for the few regions that have fossil reserves[1].

The renewable resources can increase their environmental benefit if they are considered waste, like wooden boxes at the end of life, sawdust, wasted pallets, and so on. Bioenergy production industries have a large number of production process inputs and outputs in the form of energy and materials. These material and energy flows can be used in subsequent processing steps internally or at external industries. By linking industries together to create **physical synergies** based on flows, improvements in environmental performance can be accomplished (Bistagnino, 2009). By-products synergies allow for the employment of products created during a production process. These by-products can be used as raw materials in subsequent processes, additives or other purposes depending upon the need and quality. In bioenergy production, many of the by-products include organic material can be used further (Martin and Eklund, 2010).

The waste-to-energy strategy has the following strenghts:
- reduction of weight and volume of the "waste mountain";
- energy recovery and replacement af fossil fuels;
- environmental control and minimization of green gasses.

However, **most reusable and marketable residuals have to be considered under the restrictive waste law,** regardless of whether they are actually disposed of or reused. Therefore, before running into the bureaucracy of waste treatment plant approval in setting up a recycling network, companies often prefer to use virgin input materials instead. The existing legislation is based upon the fact that industrial process waste is something with little value, compared to the product, or that has been strongly polluted, so much so that manufacturers regard it as an issue to be solved as quickly and cheaply as possible. From this perspective it is easy to understand why the legislator wants to protect both the environment and the people, by means of obligatory regulations that aim to chart the course performed by substances regarded as hazardous. Instead, if output is turned from a problem into a resource, producing in such way an interesting economic value, it would consequently be considered an active part of a process. As such its intrinsic qualities should be enhanced during the early processes, when it should not be downgraded. If its appealing properties are unchanged, it does not lose its economic trading value, achieving at the same time a formidable result, that being a **zero emissions productive culture**.

The way toward a massive use of bioenergy is closely linked to the transition from highly centralised energy systems to a greater share of **distributed energy technologies** (Vaitheeswaran,

---

[1] The oil reserves in the World, proved at the end 2009, are held by Middle East companies (56.6%), and by South-Central America (14.9%); the other continents and countries have lower percentages. The natural gas reserves in the World, proved at the end 2009, are held by Middle East companies (76.17%), and by Russian Federation (23.7%); the other continents and countries have lower percentages. The coal reserves in the World, proved at the end 2009, are held by USA (28.9%), by Russian Federation (19%), by China (13.9%); the other continents and countries have lower percentages. Source: BP Statistical Review of World Energy, June 2010.

2005). System resilience and energy security can be greatly increased with **distributed and dispersed power generation**. Additionally, there is an immense opportunity for organizations to network with one another in order to take advantage of more cyclical and systemic use and re-usage of energy resources. Efficient use of material resources involves planning systems that allow a single resource to serve multiple functions, and reduce need for additional inputs (Campagnaro, 2008). The combination of energy resources to attain the optimal energy supply may vary for different territorial regions to suit the local situation, which is influenced by local practices, current resource level and variety and environment. Governments need to recognize that energy and climate policy is a dynamic and changing field, and there is no one size fit all solution when deciding on interventions to reduce emissions and increase energy use efficiency. Clearly, the territorial approach encompasses the benefits of economy-wide, sectoral and local approaches, and more. Incorporating local culture allows people to feel a sense of ownership and identify with a policy intervention, and mobilizes them to contribute and participate. The territorial approach encourages **participation of people** from all segments of society and across social economical groups. A truly sustainable development may be achieved with the **diversification and localization of energy sources** and **systems** if the adverse impact of each energy system is sufficiently small and well fit within the tolerance limit of the environment. The key to success of a **decentralized, more integrated structure** is the ability to share into information or knowledge. Improving networks that increase access to knowledge in the area of bioenergy, processing, marketing, and consumption can have a huge effect on what consumers' demands and how farmers and food processing companies produce for the end market. Since much of the knowledge about energy and material networks are held in local populations, it is essential for policy decisions draw from these networks, for those from or working in these communities to disseminate local knowledge further to networks at different levels. This will help build synergistic linkages between local populations who hold traditional knowledge and experts that hold knowledge in natural sciences. It has been shown that interventions or technologies that incorporate traditional knowledge are more readily adopted. However, care must be taken to balance this marriage of new and old, and local stakeholders need to play an active role in the transformation process. The local knowledge should be shared at the regional, national and international level. This new productive model generates new businesses, offering great potential for the creation of new jobs but also implying structural change and transformation of existing jobs. In systems related to bioenergy production, the new generated jobs are green jobs, as defined by United Nations Environmental Programme (UNEP), ILO, International Organisational of Employers (IOE), and International Trade Union Confederation (ITUC), in 2008, because they reduce the environmental impact of enterprises and economic sectors[2].

---

[2] The Green Jobs Report defines green jobs as work in agriculture, industry, services and administration that contributes to preserving or restoring the quality of the environment while also meeting requirements of decent work (adequate wages, safe conditions, workers rights, social dialogue and social protection).

Meeting skills needs is a critical factor for productivity, employment growth and development. Today, **skills gaps** are already recognised as a major bottleneck in a number of sectors, such as renewable energy and manufacturing (Martinez-Fernandez, Hinojosa, & Miranda, 2010). The imposition of stricter climate change regulation will inevitably lead to significant job losses and increasing social fragmentation if appropriate steps are not taken to avoid this. That is a further reason to take in consideration the designed framework by this thesis work; it anticipates the skills to ensure the provision correspondence to current and future labour market demand for green collar workers both quantitatively and qualitatively and at different levels (i.e. national, sectoral, regional, company, training provider). The **wide knowledge about environmental issue** and the **ability to see solutions** instead of problems lead to a common understanding from politicians, businesses, inhabitants and organizations. **Participative design** is not only a **high index of democracy** but also a **secure way of success** in these complex systems that involve many actors. Thus, municipalities are carried on the responsibility to involve also the population in the decision processes and take care of its physical and moral health. As demonstrating by the findings of this research, the local governance can guide the economic development into an alternative development pathway that is pulled by innovation in social and managerial domains. It is possible because local bodies can assess the locally available, but currently under utilized, resources and steer the development under their jurisdictions in a direction that can add more value to such resources.

CONCLUSIONS

7.1 Reflections

Deregulation and liberization have profoundly changed the landscape of power generation, enhancing the role of private actors. As a result of the regulations having evolved at EU level, electricity sector has become predominantly commercial and the market players have made their investments. Changing supply conditions and the regulatory framework directly impact on power generation: in that changing scenario the designed framework has huge potentiality. In particular for the crucial aspect of the security, because energy is a pubblic good and the public authorities bear a responsibility for a market design that is conductive to ensuring that sufficient power will be on offer in order to meet future demand. In other words, private actors will make the necessary investments but public authorities are ultimately responsible for a market design that fosters energy security and encourages investment. The framework helps all actors to clearly understand their role in the development of local economies focus on production of green energy. The definition of roles and responsabilities of different actors is not reductive, but it contribute to coordinate the participative design of the complex systems. The coordination does not necessarily result in collaboration, but for sure the collaboration is done with coordiantion among parties. The cooperation is vital for the integration among companies and for the effectiveness of the projects. The actors stressed the importance of everybody to gain something from the cooperation. Communication is more important for success than technology, so managers establish acquaintance and companies hace non -secretive management style. The contractural relationships are also very important: the contracts should be detailed regarding the interaction of firms and flows of materials.

As a mechanism of governance, partnership redefines the role and domain of political-administrative authorities and private actors in the delivery of welfare services as well as community development. Rather than relying on strategies of hierarchical steering, imposition and inducements, all actors should pursue a policy of negotiation and cooperation. The idea of partnership increase the efficiency and effectiveness of the projects. The participatory perspective on the implementation of development-oriented partnership raises the question of designing more appropriate and efficient instruments for handling collective problems. Relational contracts are based on trustful negotiation and dialogue. The trust is an essential factor for inter-organisational relations that influence the development of SD projects. Most connections are governed by informal agreements, founded on mutual trust. Such governance mechanism, although including some levels of inherent risk, eliminate the need for more resource intensive governance mechanism and thereby reduce the transaction costs. In complex projects, the partecipative design is the only way to face and solve:

- corporate and social pressures;

- governmental/regulatory barriers;

- lack of tools to measure progress.

Regarding LED, the organizational umbrellas that express and regulate the partnerships should be autonomous and formally decoupled from local government. However, the fact that private actors may propose initives, does not mean that the political agenda has excluded issues of economic development. Throught participation, governmental actors may acquire an understanding of the private sector way of organizing and conducting activities; the business leaders may become more attentive to issues of community development (Harding, 1998). This thesis work shows how partnership contribute to an expansion of the agenda and a concern for more long term aim. The more companies are advanced with their environmental work, the more inclined they are to take part in SD networks. With SD methodology three different network can be designed and implemented:

- **knowledge network**: Since much of the knowledge are held in local populations, it is essential to draw from these networks. This will help build synergistic linkages between local populations who hold traditional knowledge and experts that hold knowledge in natural sciences. It has been shown that interventions or technologies that incorporate traditional knowledge are more readily adopted. However, care must be taken to balance this marriage of new and old, and local stakeholders need to play an active role in the transformation process. Networks constitute important channels for the transfer of both tacit and explicit knowledge. These networks may be built around markets and may therefore be conducive to the identification of problems and the development of new technical solutions. The local knowledge should be shared at the regional, national and international level. As information become more accessible through the global information portal, which is rapidly expanding trough increased use of Internet and telecommunications improvements, the way in which organizations govern and grow is changing. New, inexpensive opportunities to connect and share not only ideas but also business functions are expanding rapidly. Organizations are quickly adopting a more decentralized organizational structure in which greater innovation and growth take place at all sites of an organization. Both formal and informal personal and business networks are on the rise, and should be considered with equal importance. The key to success of a decentralized, more integrated structure is the ability to share into information or knowledge. Greater attention and support via knowledge networks is needed to spread a more systemic form of business and organizational management.

- **material network**: future costs of access and consumption of material resources remain too inexpensive as the environmental opportunity costs have yet to be fully quantified and spread across all units of material resources used. Given the increasing evidence of environmental degradation and climate change, it seems clear that the real costs of the vast material resources consumed and transformed globally on a daily basis have not been sufficiently accounted for. The availability of the resources is constantly changing due to human activities and environmental effects. Efficient use of material resources involves planning systems that allow a single resource to serve multiple functions, and reduce need for additional inputs (Campagnaro, 2008). An ideal system generates positive feedback, where one or more of the end products or what is perceived as "waste" becomes a new material resource that could fuel further generation of material resources. Many of these practices require little capital and modification for system-wide adoption. There is an immense opportunity for organizations to network with one another in order to take advantage of more systemic use and re-usage of materials. There is a growing practice of linking business units or enterprises to take advantage.

- **energy network**: distributed energy systems are small-scale, site-appropriate and resilient power generation facilities. Bioenergy produced in small plants and distributed in the territory helps the success and the sustainability. Such agile system can be a new paradigm for both energy efficiency and reliability for any region or country. Using these networks, energy systems can be collectively innovated, so they are beneficial for the people the environment, and pleasurable for all. With territorial nets of energy, the areas become self-sufficient at energy point of view, increasing the businesses, the job places and reducing the waste and the emissions in the environment. In that way the regions appear as a patchwork of small sustainable units that can easily change when the initial conditions are not the same.

The cross analysis of these three networks shows the caracteristics of a specific territory (*figure 7.1.I*).

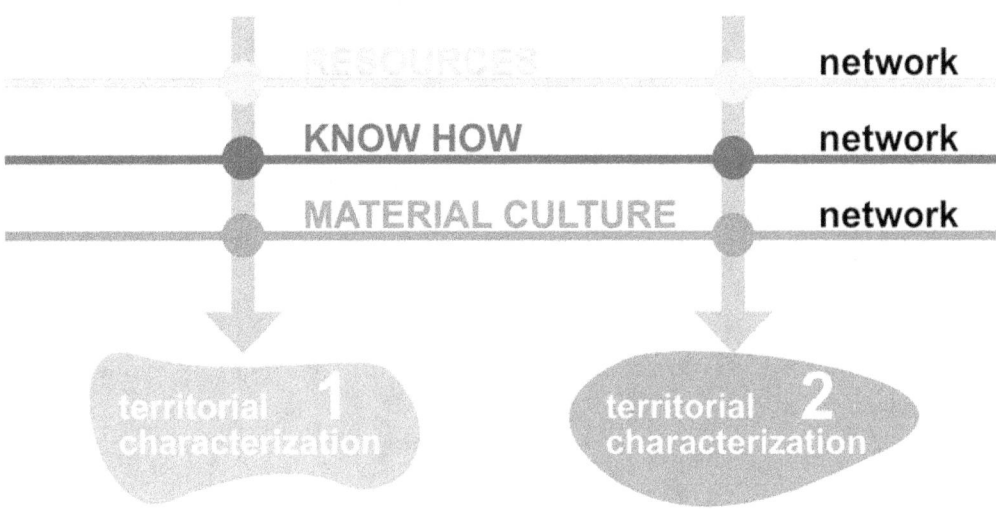

figure 7.1.I: territorial characterization through networks. *Source:* Bistagnino, L. (2011) *Systemic Design*. Bra (CN), Italy: Slow Food Editore.

When the linkages between materials, energy, people and their knowledge are mapped out clearly, efficient pathways towards sustainable ways to use and re-use untapped resources will become apparent. The criticalities shown in the first part of this thesis work can now find some answers and reflections:

- **Economic conditions**: with the new energy model proposed the economic availability is guarantee not by policy measures or subsidies, but by the cooperation among private companies that work together because it is profitable. In that way the market is not distorted external factors, furthermore the territorial nets of energy can be realized without the support of EU programme but just with the private fundindings that look to new profitable activities.

- **Know-how and institutional capacity**: The association of energy networks with know-how networks contributes to share the information and the experience of all actors involved in complex systems. The required know-how is developed in existing actors through traditional learning processes and through the participation of new actors. Learning processes and altering perceptions of public and private about renewable energy is necessary to build up legitimacy for these territorial nets of energy. In that way, the bioenergy biases of obsolete technology and of waste relation can be faced and get over. The territorial nets of energy create small units affected of PIMBY (Please In My Back Yard), instead of NIMBY. This phenomenon occurs when the power generation is absolutely safe for the health and for the environment and when it is a source of income.

- **Supply chain coordination**: the territorial nets of energy are designed in a participative way and grew up in a cooperative way. The transparency and trust are the basis for the relations among to carry out successful projects during the various stage of the overall project lifetime. The actors come from the same territory so they understand and rule better each others. These nets bridge distances among partners, that have different interests and motives, but common and complex project.

In the end, the territorial nets of energy generate:

- Distributed and diversified energy systems;

- small and flexible power generations;

- energy safe services;

- active management where also citizens are participative in production, consumption, management,...

- internalised cost.

7.2 Contributions

It is held that this research can be seen as having contributed to the maturation of SD in the bioenergy field. This is done through strenghtening the theoretical claims via provisions of empirical evidence, and designing a new tested framework.

This thesis work contributes to knowledge on bioenergy and it has value for the research community, policy-makers (and/or policy advisors), local municipalities, and industrial actors. While the contribution merge into each other, they can be classified under four themes:

- **fieldworks**: this thesis work has been based on extensive fieldwork and test on real case. Analysing eight case studies has involved face-to-face interviews and discussions with a range of actors (such as representative from local municipalities, plant owners, employees, academics, and farmers) and site visits to various locations relevant to bioenergy systems, in particular in Sweden. The combined data collection and analysis activities represent an empirical contribution to the body of knowledge on bioenergy systems.

- **methodology**: this thesis is based on a combination of desk and field research. Literature reviews, case studies, site visits, stakeholder interviews, industry interactions, and research are used to design and test the framework to develop local economies based on bioenergy. This represent an attempt to apply a more robust research strategy to explore implementation of bioenergy systems and to go down the SD in that field. The research process involved the participation in the development of Agrindustria project thereby directly influencing actors in the field, and adding another dimension to the research work.

- **Knowledge**: this research work responds to calls in the literature and scientific circles for a greater understanding and analysis of non-technical issue surrounding bioenergy systems. The focus of social, environmental, and economical aspects of bioenergy systems can therefore be considered a rather novel approach for this field. This encourage a more holistic perspective with the use of knowledge from different disciplines and with the maps of interactions between elements in bioenergy systems. Stepping back and looking at both the technical aspects and overarching social and economical aspetcs of bioenergy systems opens up "new" insights that are particularly relevant for a better understanding of implementation challenges.

- **Diffusion**: the findings of this research were presented at a number of conferences with international audiences and also in informal meetings. Research may be of little or no value to actors if the findings are not made directly available to them. Written and formal information (peer-reviewed scientific journal, the essay on books, the proceedings and the new modes of scholarly communication) are integrated with personal and informal information, in which actors whose technicaland analytical capacity can actively partecipate.

REFERENCES

Allenby, B.R. (1994). Industrial Ecology gets down to the Earth: addressing the New Challenges of Practical Engineering in an Environmentally Constrained World. *Circuit & Devices Magazine, 10 (1),* 24-28.

Alvesson, M., Sköldenberg, K. (2000). *Reflexive Methodology: New Vistas For Qualitative Research.* London, England: Sage Publications.

Berndes, G., Hoogkijk, M., Van Den Broek, R. (2003). The Contrinution of biomass in the future global supplì: a review of 17 studies. *Biomass and Bioenergy, 25,* 1-28.

Barbero, S. (2012). *Systemic Energy Network. The Practice of macro and micro case studies.* Raleigh N.C., USA: Lulu Enterprises, Inc.

Bistagnino, L. (2009). *Design Sistemico. Progettare la sostenibilità produttiva e ambientale.* Bra (CN), Italy: Slow Food Editore.

Bloor, M. (1997). Techniques of Validation in Qualitative Research: a Critical Commentary. In G. Miller & R. Dingwall (Eds.), *Context and Method in Qualitative Research* (pp. 37-50). London, England: Sage Pubblication.

Boyle, G. (2004). *Renewable Energy.* Oxford, England: Oxford University Press.

Bp, (2010). *Statistical Review Of World Energy.* London, England: Beacon Press.

Bryman, A. (2004). *Social Research Methods.* Oxford, England: Oxford University Press.

Calderini, M., Garrone, P., & Sobrero, M. (2003). *Corporate Governance, Market Structure and Innovation.* Cheltenham, England: Edward Elgar Publishing Limited,

Campagnaro, C. (2008). *Cinque Miliardi di Sfere.* Cesano Boscone (MI), Italy: Casa Editrice Ambrosiana Zanichelli.

Campbell, S. (1996). Green cities, growing cities, just cities? Urban planning and the contradictions of sustainable development. *Journal of the American Planning Association, 62 (3),* 296-312.

Capra, F. (1997). *The Web of Life: a New Synthesis of Mind and Matter.* London, England: Flamingo.

Capra, F. (2002). *The hidden connections: A science for sustainable living.* New York, NY: Doubleday.

Celaschi, F., & Deserti, A. (2007). *Design e Innovazione. Strumenti e Pratiche per la ricerca apllicata.* Roma, Italy: Carocci Editore.

Chertow, M.R. (2000). Industrial Symbiosis: Literature and Taxonomy. *Annual Review of Energy and Environment, 25,* 313-337.

Chertow, M.R. (2007). Uncovering Industrial Symbiosis. *Journal of Industrial Ecology, 11 (1),* 11-30.

Chertow, M. R., Ashton, W., & Kuppali, R. (2004). *The Industrial Symbiosis Research Symposium at Yale: Advancing the Study of Industry and Environment.* New Haven, CT: Yale School of Forestry & Environmental Studies.

Commission of The Europea Communities, (1997). *White Paper For A Community Strategy And Action Plan on Energy For The Future And Renewable Sources Of Energy*, Com (97)599. Brussels, Belgium: Eu Commission.

Commission of The Europea Communities, (2005). *Biomass Action Plan.* Brussels, Belgium: Eu Commission.

Committee on Employment And Social Policy (2008). *Employment And Labour Market Implications Of Climate Change*, Gb. 303/Esp/4. Geneva, Switzerland: International Labour Office.

Connell, R.W. (1985). *How To Supervise A Phd.* Sidney, Australia: Vestes.

Cosgriff Dunn, B., & Stainemann, A. (1998). Industrial Ecology For Sustainable Communitie. *Journal of Environmental Planning and Management, 41 (6),* 661-672.

Daly, H.E. (1996). *Beyond Growth: the Economics of Sustainable Development.* Boston, MA: Beacon Press.

Desrochers, P. (2004), Industrial Symbiosis: the case for market coordination. *Journal of Cleaner Production*, 12 (8-10), 1099-1110.

Deutz, P., & Gibbs, D. (2008). Industrial Ecology And Regional Development: Eco-Industrial Development As Cluster Policy. *Regional Studies, 42 (10),* 1295-1298.

Directive Of European Parliament & Council (2001). *The Promotion Of The Electricity Produced From Renewable Energy Sources In The Internal Electricity Market, 2001/77/Ec.* Luxembourg, Luxembourg: Official Journal Of The European Communities.

Directive Of European Parliament & Council (2003). *A Scheme For Green House Gas Emissions*

Allowance Trading Within The Community, 2003/87/Ec. Luxembourg, Luxembourg: Official Journal Of The European Communities.

Donolo, C. (2001). *Disordine.* Roma, Italy: Donzelli.

Ehrenfeld, J.R. (1997). A Framework for Industrial Ecology. *Journal of Cleaning Production, 5 (1-2),* 87-95.

Faaij, A.P.C. (2006). Emerging International Bio-Energy Markets And Opportunities For Socio-Economic Development. *Energy For Sustainable Development, 10 (1),* 7-19.

Fiksel, J. (2003). Design Resilient, Sustainable Systems. *Environmental Science & Technology, 37 (33),* 5330-5339.

Fiksel, J. (2006). Sustainability And Resilience: Toward A Systems Approach. *Sustainability: Science, Practice & Policy, 2 (2),* 14-21.

Frosch, R.A. (1992). Industrial Ecology: a Philosophical Introduction. *Proceedings of the National Academy of Scienc, 89,* 800-803.

Frosh R.A., & Gallopoulos, N.E. (1989). Strategies for Manufacturing. *Scientific American, 3 (189),* 94-102.

Geller, H. (2003). *Energy Revolution: Policies For Sustainable Future.* Washington, Wa: Insland Press.

Geels, F.W. (2004). From Sectoral Systems of Innovation to Socio-Technical Systems: insight about dynamics and change from sociology and institutional theory. *Research Policy, 33,* 897-920.

Germak, C. (2008). *Man at the centre of the project. Design for a New Humanism.* Turin, Italy: Allemandi & C.

Goleman, D. (2009). *Ecological Intelligence: how knowing the hidden impacts of what we buy can change everything.* New York, NY: Crown Business.

Hall, D., & Scrase, J. (1998). Will Biomass Be The Environmentally Friendly Fuel Of The Future? *Biomass and Bioenergy, 15 (4-5),* 357-367.

Hammersley, M. (2000). The Relevance of Qualitative Research. *Oxford Review of Education, 26 (3-4),* 393-405.

Harding, A. (1998). Public-Private Partnership in the UK. In J. Pierre (Ed.), *Partnership In Urban*

Governance. European And American Experiences. Basingstoke, Ny: Palgrave.

Hardy, C., & Graedel, T. (2002). Industrial Ecosystems as Food Webs. *Journal of Industrial Ecology, 6 (1),* 29-38.

Heylinger, F., Joslyn, C., & Turchin, V. (2000). *Principia Cybernetica.* http://pespmc1.vub.ac.be.

Italian Ministry of Economic Development (2009). *Italian National Renewable Energy Action Plan.* Directive 2009/28/Ec and Commission Decision of 30 June 2009. Brussels, Belgium: Eu Commission.

Johansson, B. (2008). *Bioenergy – For What And How Much?* Stockholm, Sweden: Forskningsrådet Formas.

Johansson, A., Kisch, P., Mirata, M. (2004). Distributed Anomie – A New Engine for Innovation. *Journal of Cleaner Production, 13,* 971-979.

Kant, I. (1785), *Grundlegung Zur Metaphysik Der Sitten.* Riga, Germany: Johann Friedrich Hartknoch

Krampen, M., & Hörmann, G. (2003). *The Ulm School Of Design: Beginnings Of A Project Of Unyielding Modernity.* Berlin, Germany: Ernst & Sohn.

Lanzavecchia, C. (2004). *Il fare ecologico.* Turin, Italy: Time&Mind.

Ligabò, G. (2007). *L'ambiente e L'energia da Fonti Rinnovabili.* Reggio Emilia, Italy: Edizioni Diabasis.

Lovelock, J. E. (1988). *The Ages of Gaia: a Biography of our Living Earth.* New York, NY: Norton.

Lovins, A. B., & Hawken, P., & Lovins, L. H. (1999). *Natural capitalism: Creating the next industrial revolution.* New York, NY: Little Brown&Company.

Lowe, E. A. (1997). Creating by-product resource exchanges for Eco-Industrial Parks. *Journal of Cleaner Production, 5 (1-2), Industrial Ecology Special Issue,* 57-66.

Lund, H. (2006). Renewable Energy Strategies for Sustainable Development. *Sciencedirect, Energy, 32,* 912-912.

Mandelbrot, B. B. (1982). *The Fractal Geometry of the Nature.* New York, NY: W.H. Freeman and Company.

Maturana, H. R., & Varela, F.J. (1972). *De Maquinas y Seres Vivos. Una teoria sobra la organizacion biologica.* Santiago de Chile, Chile: Editorial Universitaria.

Marshall, A. (1920). *Principles of Economics.* London, England: Macmillan.

Martin, M., & Eklund, M. (2010). Improving The Environmental Performance of Biofuels with Industrial Symbiosis, *Biomass and Bioenergy, forthcoming.*

Martinez-Fernandez, C., Hinojosa, C., & Miranda, G. (2010). *Greening Jobs And Skills: Labour Market Implications Of Addressing Climate Change.* Paris, France: Oecd Local Economic And Employment Development Working Paper Series.

McCormick, K. (2007). *Advancing Bioenergy in Europe. Exploring bioenergy systems and socio-political issues* (Published doctoral dissertation). The International Institute for Industrial Environmental Economics, Lund, Sweden.

Mccormick, K., & Kåberger, T. (2005). Exploring A Pioneering Bioenergy System: The Case Of Enköping In Sweden. *Journal of Cleaner Production, 13,* 1003-1014.

Mcculloch, W.S., & Pitts, W.H. (1948). A Logical Calculus Of The Ideas Immanent In Nervous Activity. *Bulletin Of Mathematical Biophisics, 5,* 115-133.

Meadows, D. (1999). *Leverage points: places to intervene in a System.* Hartland, WI: Sustainability Institute.

Mirata, M. (2005). *Industrial Symbiosis. A tool for more sustainable regions?* (Published doctoral dissertation). The International Institute for Industrial Environmental Economics, Lund, Sweden.

Morrow, R.A., & Brown, D.D. (1994). *Critical Theory and Methodology.* London, Engrand: Sage Publications.

Moslehi, K., & Kumar, R. (2010). Smart Grid – A Reliability Perspective. *Innovative Smart Grid Technologies.* Washington DC, WA: IEEE PES.

Nussbaum, M. C. (2000). *Women and Human Development: the Capabilities Approach.* Cambridge, England: Cambridge University Press.

Openshaw, D. (1998). Embedded Generation in Distribution Networks: with experience of UK commerce in electricity, *Wind Engineering, 22 (4),* 189-196.

Ottosson, S. (2003). Participation Action Research: a Key to improve knowledge of management. *Thechnovation, 23,* 87-94.

Pezzoli, K. (1997). Sustainable Development: a Transdisciplinary Overview of the Literature. *Journal of Environmental Planning & Management, 40 (5)*, 549-574.

Pauli, G. (2010). *The Blue Economy: 100 Innovations inspired by nature that can generate over a decade 100 million jobs.* Taos, NM: Paradigm Publications.

Pisek, P.E., & Wilson, T. (2001). Complexity, Leadership, And Management In Healthcare Organizations. *British Medical Journal, 323,* 746-749.

Porter, M.E. (1990). *Competitive Advantage of Nations.* New York, Ny:Free Press.

Porter, M.E. (1998). *On Competition.* Boston, Ma: Harward Business School Press.

Porter, M.E. (2000). Localization, Competition, Ad Economic Development: Local Clusters In A Global Economy. *Economic Development Quarterly, 14 (1),* 15-35.

Regeringskansliet (2009). *The Swedish National Action Plan for the Promotion of the use of Renewable Energy.* Directive 2009/28/Ec and Commission Decision of 30 June 2009. Brussels, Belgium: Eu Commission.

Rifkin, J. (2002). *The Hydrogen Economy.* New York, NY: Tarcher.

Rifkin, J. (2009). *The Empathic Civilization: The Race To Global Consciousness in a World In Crisis.* New York, NY: Tarcher.

Roos, A., Graham, R. L., Hektor, B., & Rakos, C. (1999). Critical Factors To Bioenergy Implementation. *Biomass and Bioenergy, 17,* 113-126.

Rösch, C., & Kaltschmitt, M. (1999). Energy From Biomass – Do Non-Technical Barriers Prevent An Increased Use? *Biomass and Bioenergy, 16,* 347-356.

Shrödinger, E. (1946). *What is Life? The physical aspects of living cells.* Cambridge, England: Cambridge University Press, The Macmillan Company.

Sen, A. (1979). *Equality of What? The Tanner Lecture on Human Values.* Stanford, CA: Stanfird University Press.

Senge R. E. (1993). *The Art of Case Study Research,* Thousand Oaks, CA: Sage Pubblication.

Sgreccia, E. (2007). *Manuale Di Bioetica vol.1: fondamenti ed etica biomedica.* Milan, Italy: Vita e Pensiero.

Shannon, C.E. (1948). A Mathematical Theory Of Communication. *Bell System Thecnical Journal, 27 (623-656)*, 379-423.

Socolow, R.H., Andrews, C., Berkhout, F., & Thomas, V. (1994). *Industrial Ecology and Global Change*. Cambridge, England: Cambridge University Press.

Star, S.L., & Griesemer, J. (1989). Institutional Ecology, "Translation" and Boundary Objects: Amateurs and Professionals in Berkeley's Museum of Vertebrate Zoology, 1907-39. *Social Studies and Sciencie, 19 (3)*, 387-420.

Tibbs, H. C. (1993). *Industrial Ecology: an Environmental Agenda for Industry*. Emeryville, CA: Global Business Network.

Unep, Ilo, Ioe, & Ituc (2008). *Green Jobs: Towards Decent Work In A Sustainable, Low-Carbon World*. Nairobi, Kenya: Division Of Communications And Public Information.

Vaitheeswaran, V. (2005). *Power and People*. London, England: Earthscan.

Van Berkel, R. (2006). Regional Resource Synergies for Sustainable Developmenti Heavy Industrial Area: an Overview of Opportunity and Experiences. *Cleaner Production, Bulletin of Curtin University of Technology, 1,* 1-139.

Von Bertalanffy, L. (1969). *General System Theory: Foundations, Development, Applications.* New York, NY: George Braziller, Inc.

Wiener, N. (1948). *Cybernetics: or Control and Communication in the Animal and the Machine*. Paris, France: Hermann & Cie.

Wolf, A. (2007). *Industrial Symbiosis in the Swedish Forest Industry* (Published doctoral dissertation). Division of Energy Systems, Linköping, Sweden.

World Energy Assessment, (2000). E*nergy And The Challenge Of Sustainability*. New York, NY: United Nation Development Programme.

World Energy Assessment (2004). *Overview Update*. New York, NY: United Nations Development Programme.

www.ingramcontent.com/pod-product-compliance
Lightning Source LLC
Chambersburg PA
CBHW080922170526
45158CB00008B/2200